T0256210

Non-Destructive Diagnostics of Concrete Floors

Concrete floors should be diagnosed in order to obtain the proper durability. Non-destructive testing (NDT) methods, which have numerous advantages and are very effective for *in situ* testing, are recommended for this purpose. *Non-Destructive Diagnostics of Concrete Floors: Methods and Case Studies* offers useful NDT methods, test methodologies, and case studies. This book contains classifications of NDT methods, examines their areas of usefulness in floor diagnostics, and explains the complementarity and reliability of NDT methods as well as the need to calibrate research equipment. It presents interesting case studies of concrete floors, such as dowelled floors, floors with a top layer made of stone slabs, industrial floors, industrial floors with a top layer of polyurethane-cement, layered floors, post-tensioned floors, and cement screeds. The authors have drawn on many years of experience in both academia and the practical diagnosis of concrete floors using NDT methods.

Łukasz Sadowski is an Associate Professor and Head of the Department at the Faculty of Civil Engineering of Wroclaw University of Science Technology, Poland.

Professor Jerzy Hoła is an employee and former dean of the Faculty of Civil Engineering of Wroclaw University of Science and Technology, Poland.

Non-Destructive Diagnostics of Concrete Floors

Methods and Case Studies

Łukasz Sadowski
Jerzy Hoła

CRC Press
Taylor & Francis Group
Boca Raton London New York

CRC Press is an imprint of the
Taylor & Francis Group, an **informa** business

First edition published 2023
by CRC Press
6000 Broken Sound Parkway NW, Suite 300, Boca Raton, FL 33487-2742

and by CRC Press
4 Park Square, Milton Park, Abingdon, Oxon, OX14 4RN

CRC Press is an imprint of Taylor & Francis Group, LLC

© 2023 Łukasz Sadowski and Jerzy Hoła

ISBN: 978-1-032-26452-3 (hbk)
ISBN: 978-1-032-26453-0 (pbk)
ISBN: 978-1-003-28837-4 (ebk)

DOI: 10.1201/9781003288374

Typeset in Times
by KnowledgeWorks Global Ltd.

Contents

Preface

This book is devoted to non-destructive diagnostics of concrete floors. It contains new knowledge and research experience, which is preceded by an update and systemization of the already available knowledge concerning the subject covered in the title. This book meets the needs of those who already deal with, or intend to deal with, the assessment of the condition of floors, as well as those who seek support in literature with regards to, among others, the field of non-destructive testing methods, methodologies, and case studies.

This book is divided into 14 chapters. Chapter 1 provides an introduction. Chapter 2 is devoted to the problem of the degradation of floors. Chapter 3 provides introductory information to the diagnostics of floors. Chapter 4 presents the classifications of non-destructive and semi-destructive methods, and indicates the areas of their usefulness in the diagnostics of floors. In turn, Chapter 5 contains synthetic descriptions of selected non-destructive and semi-destructive methods useful in floor diagnostics. Chapter 6 is devoted to the reliability of the use of non-destructive methods, while Chapter 7 deals with the issues of the calibration and scaling of research equipment. Chapters 8–14 contain case studies concerning, in order of diagnostics: dowelled concrete floors, concrete floors with a top layer made of stone, industrial concrete floors, concrete floors with a top layer made of polyurethane, cement screeds, layered concrete floors, and post-tensioned concrete floors. For researchers and experts who want to broaden their knowledge with regards to the subject of this book, useful reference items are provided at the end of each chapter.

While writing this book, it was not only literature studies that were used, but also the authors' own thoughts and many years of experience acquired in both academia and in the practical diagnostics of floors using non-destructive methods. The authors of this book are recognized practicing specialists in this field, and they are also, or have been, active participants in scientific committees and scientific and technical events related to non-destructive testing in the construction industry.

In the opinion of the authors the published book should not only meet the expectations of experienced scientists and experts specializing in the field of non-destructive testing in floors, but also civil engineers, both experienced and beginners, in the field of the non-destructive diagnostics of floors. Moreover, students of construction and architecture courses should consider this book a useful source of knowledge in the field of non-destructive and semi-destructive methods of testing and diagnosing floors.

Authors

Łukasz Sadowski is an Associate Professor at the Faculty of Civil Engineering of Wroclaw University of Science Technology. He is the Head of the Department of Materials Engineering and Construction Processes. He was a fellow of the Foundation for Polish Science and the Ministry of Science and Higher Education. He is a senior member of RILEM and a member of the Polish Academy of Sciences. His achievements include more than 200 publications, 2 books, and tens of expertise reports.

Professor Jerzy Hoła is employed at the Faculty of Civil Engineering of Wroclaw University of Science and Technology (WUST). He is a former Dean of the Faculty and a former senator of WUST. He is a member of the Civil Engineering Committee and the Chairman of the Construction and Mechanics Committee of the Polish Academy of Sciences. He specializes in general construction and non-destructive testing. He has published more than 300 scientific works, including several books and several hundred expertise reports.

1 Introduction

One should agree with the statement that floors undergo degradation during operation, resulting in a gradual reduction in their durability [1–5]. In various operational situations, the intensity of these processes may be different in the sense that it may be slowed down or accelerated [6–8]. It can be slowed down when floors are properly protected against the adverse effects of environmental factors, are free from defects, and are properly used and maintained. It can be accelerated, in particular, as a result of the action of various extreme weather conditions or mechanical factors (single or cumulative), due to the presence of significant workmanship or design defects, due to improper use, and also due to negligence in maintaining a proper technical condition. The consequence of degradation processes is the occurrence of more serious or less serious damage in the floor, which in turn requires repair. In some cases, it may be necessary to strengthen the damaged floor, for example, by over-concreting, without which more serious damage may occur, resulting in the floor being excluded from operation.

When taking care of the original durability and safety of use of floors, they should be subjected to preventive technical inspections. These inspections, as a desirable element of proper maintenance, are also performed in necessary situations, such as a sudden deterioration of the technical condition. Such a situation may occur, for example, as a result of damage caused by the disclosure of an existing significant latent defect in the floor or in the floor's sub-base, due to extreme weather conditions, or due to overload or fire in the object. When a structure is in use, there are more serious defects and irregularities, or the defects that are revealed usually result in the need to conduct a diagnostic procedure. Conducting such a procedure may also result from the need to assess the possibilities and conditions of the planned modernization, or operational changes resulting in a change in the nature and value of the current loads.

Diagnostic management is very important, the results of which constitute the basis for the preparation of an appropriate technical study. Such a study is, for example, an expert study, which is usually characterized by a high degree of detail and inquisitiveness. For this reason, in diagnostic procedures, it is even necessary to use a variety of research methods and techniques that are appropriately selected for a given situation and needs.

Among the many available research methods and techniques that are useful for diagnosing floors, non-destructive methods and techniques deserve special attention. This is mainly due to their numerous advantages, such as their very high usefulness in in situ studies, the fact that there is no interference in the structure of the tested floor, and the possibility of multiple tests in any number of places and at any time. It is worth noting at this point that some of the non-destructive methods, such as sclerometric or ultrasonic methods, have long been standardized in terms of their indirect methodical evaluation of the compressive strength of concrete in a floor [9–11]. In addition to non-destructive methods, in many situations it is also advisable to use non-destructive methods that do not cause significant violations of the structure of

DOI: 10.1201/9781003288374-1

the tested floor. Moreover, in the case of these methods, some of them are standardized in relation to the strength tests of concrete. An example is the pull-out method [12].

Recently, significant technical progress has been noticed – on the one hand with the development of the known non-destructive methods, and on the other hand with the development of new non-destructive measurement methods and techniques. These methods are more and more often enriched with specialized software, which not only facilitates and speeds up the recording of measurement data, but is also useful in their interpretation. An example in this case is ultrasonic tomography. This development is largely aimed at creating the possibility of floor diagnostics in elements accessible from one side, which include floors. In existing floors, there is often a need to assess their thickness, locate the reinforcement, assess the depth of formed cracks, or locate various undesirable defects or damage on the surface. Moreover, in the case of newly erected floors accessible from one side, there is a need to support the effective control of their execution using non-destructive methods. Such methodical control is very advisable, especially in the case of large-area floors. This is because it favours the fulfilment of the required requirements, and minimizes or even eliminates manufacturing defects, which is important from the point of view of the durability and safety of use of floors.

Expert practice shows that it is mainly large-surface floors in which various significant defects occur during their exploitation (especially defects related to their execution). The sources of such defects can be seen to be due to the absence of an effective methodological control carried out at the stage of execution [13–16]. These are large-surface dowelled concrete floors used on port quays and in storage and manoeuvring yards; large-surface industrial floors in production and warehouse facilities; large-area concrete industrial floors with a polyurethane-cement top layer in industrial and warehouse facilities with the required increased hygiene regime; and concrete floors with a stone top layer in large-area commercial and service facilities, as well as other representative facilities. Taking into account the significant increase in the number of buildings of this type today, the problem of methodological quality control of the execution of the large-scale flat floors in them should not be underestimated [17]. The problem of significant defects occurring during operation also applies to cement floors that are commonly used in housing construction.

In relation to the above considerations, it should be said that the purpose of this book is to share knowledge and experience in the field of floor diagnostics. According to the authors, this is a desirable form of meeting the expectations of those who, for example, assess the technical condition of floors and who seek support in literature regarding research methodology or case studies. Therefore, this book introduces the problem of the degradation of floors and the need to diagnose them, and includes a classification of non-destructive and semi-destructive methods with their usefulness areas, synthetic descriptions of selected methods, and the problems concerning the reliability of using non-destructive methods in concrete floor diagnostics. This book also looks at the calibration and scaling of research equipment. It contains interesting case studies that are enriched with proprietary non-destructive testing methodologies, including those related to dowelled concrete floors, large-surface industrial floors, large-surface polyurethane-cement floors, stone floors, and layered floors. Cement screeds were also not forgotten.

The authors hope that this book will be useful to both experienced researchers and experts specializing in non-destructive testing, as well as, and even especially, engineers first encountering the problem of floor diagnostics using non-destructive methods. It should also be useful for students studying in the faculties of construction and architecture, both with regards to its content and given therein sources of reference.

REFERENCES

1. Czarnecki, L., Łukowski, P., & Garbacz, A. (2017). Repair and protection of concrete structures: Commentary to PN-EN 1504 (in Polish). Warsaw: Polish Scientific Publishers PWN SA.
2. De Schutter, G. (2012). Damage to concrete structures. Boca Raton, FL: CRC Press.
3. Sarja, A., & Vesikari, E. (Eds.). (2004). Durability design of concrete structures. Boca Raton, FL: CRC Press.
4. Dyer, T. (2014). Concrete durability. Boca Raton, FL: CRC Press.
5. Gjørv, O.E. (2009). Durability design of concrete structures in severe environments. Boca Raton, FL: CRC Press.
6. Zybura, A., Jaśniok, M., & Jaśniok, T. (2017). Diagnostics of reinforced concrete structures (in Polish). Warsaw: Polish Scientific Publishers PWN SA.
7. Bungey, J.H., Millards, S.G., & Grantham, M.G. (2006). Testing of concrete in structures. Boca Raton, FL: CRC Press.
8. Malhotra, V.M., & Carino, N.J. (2003). Handbook on nondestructive testing of concrete. Boca Raton, FL: CRC Press.
9. EN 12504-2:2013-03 Testing concrete in structures. Part 2: Non-destructive testing. Determination of the rebound number
10. EN 12504-2:2013-03 Testing concrete in structures. Part 4: Determination of ultrasonic wave velocity
11. Brunarski, L., & Dohojda, M. (2015). Diagnostyka wytrzymałości betonu w konstrukcji. Warszawa: Instytut Techniki Budowlanej.
12. EN 12504-2:2013-03 Testing concrete in structures. Part 2: Non-destructive testing. Determination of the pull-out force.
13. Sadowski, Ł., Hoła, J., Żak, A., & Chowaniec, A. (2020). Microstructural and mechanical assessment of the causes of failure of floors made of polyurethane-cement composites. Composite structures, 238, 112002.
14. Hoła, J., Sadowski, Ł., & Nowacki, A. (2019). Analysis of the causes of cracks in marble slabs in a large-surface floor of a representative commercial facility. Engineering failure analysis, 97, 1–9.
15. Hoła, J., Sadowski, Ł., & Hoła, A. (2019). The effect of failure to comply with technological and technical requirements on the condition of newly built cement mortar floors. Proceedings of the institution of mechanical engineers, part l: Journal of materials: Design and applications, 233(3), 268–275.
16. Hajduk, P. (2018). Design and technical assessment of concrete industrial floors (in Polish). Warszawa: Polish Scientific Publishers PWN.
17. Sadowski, Ł., Hoła, A., & Hoła, J. (2020). Methodology for controlling the technological process of executing floors made of cement-based materials. Materials, 13(4), 948.

2 The Problem of the Degradation of Concrete Floors

2.1 MECHANISMS OF DEGRADATION

When designing and constructing floors, it is assumed, depending on their intended use, that they will be able to operate for several dozen years without failure [1]. However, they sometimes require repair due to the appearance of more serious or less serious defects and the occurrence of damage. The damage results from the unfavourable influence of various environmental and utility factors, which contribute to the degradation of concrete and reinforcing steel that progresses over time [2–6]. To illustrate the following, the principal statements of these factors are shown schematically in Figure 2.1 (on the basis of [7–9]).

The basic mechanisms of the degradation of floors include, among others, physical, chemical, and biological mechanisms. Physical mechanisms are the most numerous groups. They are related to physical phenomena and contribute to the progress of, inter alia, such degradation mechanisms as the accumulation of inorganic dirt, cyclic freezing/thawing, erosion, salt crystallization, the impact of extreme temperatures, shrinkage and creep, overload, leaching, dampness, flooding, fatigue, changes of geotechnical conditions, hit, and abrasion. A less numerous groups can be seen to be chemical mechanisms that are associated with chemical phenomena, which contribute to the progress of, inter alia, such degradation mechanisms as carbonation, corrosion, the effects of aggressive substances, as well as the reactivity between concrete components. The least common degradation mechanisms are biological mechanisms, which include, among others, the accumulation of organic dirt, the impact of microorganisms, and the impact of plants.

2.2 CLASSIFICATION OF DEGRADATION MECHANISMS

Among the degradation mechanisms of floors, physical, chemical, and biological mechanisms should be mentioned [10–12]. These mechanisms can be divided into basic and additional mechanisms. In practice, there is usually a combination of a variety of basic and additional mechanisms. The result is the emergence of complex degradation processes that cause damage to a floor, and as a result they are decisive for its safe use, reliability, and durability. The activity and practical significance of the above-mentioned degradation mechanisms vary depending on the material from which the floor is made. Figures 2.2 and 2.3 illustrate this for concrete and steel reinforcement.

DOI: 10.1201/9781003288374-2

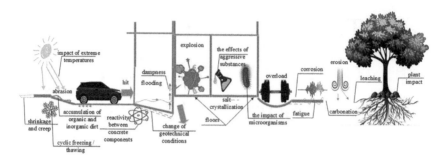

FIGURE 2.1 Diagram showing the basic mechanisms of the degradation of floors.

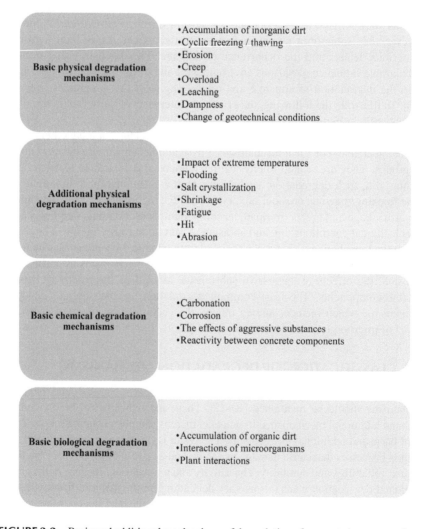

Basic physical degradation mechanisms	•Accumulation of inorganic dirt •Cyclic freezing / thawing •Erosion •Creep •Overload •Leaching •Dampness •Change of geotechnical conditions
Additional physical degradation mechanisms	•Impact of extreme temperatures •Flooding •Salt crystallization •Shrinkage •Fatigue •Hit •Abrasion
Basic chemical degradation mechanisms	•Carbonation •Corrosion •The effects of aggressive substances •Reactivity between concrete components
Basic biological degradation mechanisms	•Accumulation of organic dirt •Interactions of microorganisms •Plant interactions

FIGURE 2.2 Basic and additional mechanisms of degradation of concrete in concrete floors.

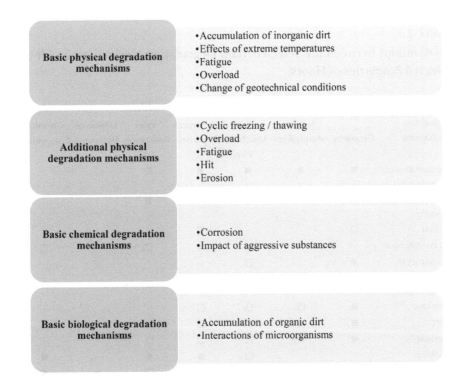

FIGURE 2.3 Basic and additional mechanisms of degradation of steel reinforcement in floors.

It is worth noting that the progression of degradation processes over time in the case of floors made of concrete is inevitable, as is also the case of floors made of other materials. However, the intensity of these processes may vary in different operating situations. They can be slowed down when a floor is correctly used and properly protected against adverse environmental influences. However, they can also accelerate, for example, as a result of sudden mechanical factors such as overload, impact or explosion, and the occurrence of extraordinary weather conditions; other random events such as extreme rainfall or fire, improper use, a lack of care for the structure over time; due to its use; and also as a result of errors made in terms of execution or design.

The consequence of degradation processes, when the primary durability of the floor becomes insufficient, is the appearance of more serious or less serious damage in it, which in turn requires repair or strengthening [13]. However, such degradation processes can also be the cause of more serious damage, which can lead to taking the floor out of service.

2.3 RELATIONSHIPS BETWEEN DEGRADATION MECHANISMS AND THE PROPERTIES OF FLOORS

As previously noted, in practice, as a rule, there are combinations of different mechanisms that result in complex degradation processes, which in turn cause damage to the floor and ultimately determine its service life. The main degradation mechanisms affecting the durability of floors are related to changes in the geometry of the

TABLE 2.1

Relationships between the Selected Main Degradation Mechanisms and the Selected Properties of Floors

Selected Main Degradation Mechanisms	Selected Properties of Floors						
	Geometry	Surface Morphology	Structure	Mechanical Properties	Water Resistance	Degree of Dampness	Chemical Composition
Physical							
Accumulation of inorganic dirt	■	■	■	■	□	-	-
Cyclic freezing/thawing	□	■	■	■	■	-	-
Erosion	□	-	-	□	-	-	-
Salt crystallization	-	-	■	■	□	-	□
The impact of extreme temperatures	■	-	□	-	-	□	-
Shrinkage	■	□	□	□	-	-	-
Creep	■	■	□	□	□	-	□
Overload	■	■	□	□	□	-	-
Leaching	-	-	□	■	■	-	■
Dampness	□	-	□	■	□	■	■
Flooding	-	-	■	■	□	■	■
Fatigue	-	□	■	■	□	-	-
Change of geotechnical conditions	■	□	□	■	□	-	-
Hit	■	□	■	■	□	-	-
Abrasion	□	■	□	□	□	-	-
Chemical							
Carbonation	-	-	-	□	□	-	■
Corrosion	-	□	■	■	□	-	■
The effects of aggressive substances	-	-	□	□	-	-	■
Reactivity between concrete components	-	□	■	■	□	-	■
Biological							
Accumulation of organic dirt	-	□	□	-	-	□	■
The impact of microorganisms	-	□	□	-	-	□	■
Plant impact	□	-	□	□	□	□	■

■ – Basic mechanism; □ – Additional mechanism; - Not applicable

floor (deformation and displacement, changes in the dimensions, etc.), changes in the morphology of the floor (the formation of unevenness, scratches), changes in the floor's structure (the formation of cracks, delamination, etc.), changes in the mechanical properties of the floor, changes in the water resistance of concrete, changes in the degree of dampness in concrete, and changes in the chemical composition of concrete. The relationships between the selected main degradation mechanisms and the selected properties of floors are presented in Table 2.1.

The inevitability of the progression of degradation processes over time, which deteriorate the condition of a floor in the sense of successively lowering its safety of use and shortening its life time, forces preventive measures to be taken. These activities include, first of all, technical inspections that differ in terms of scope and detail. These inspections are usually individualized due to diverse environmental and operational conditions, as well as the specificity of the work of individual floors. The result of these reviews are technical studies with a scope and detail tailored to the purpose they are supposed to serve. Some of these studies are based on the results of diagnostic procedures carried out with the use of various research methods [14].

2.3 CONCLUSIONS

This chapter presents the problem of the degradation of concrete floors. Attention is drawn to the unfavourable impact of various environmental and utility factors, contributing to the degradation mechanisms of concrete and steel progressing over time. These mechanisms were divided into physical, chemical, and biological, and the basic and additional ones were classified. The relationships between these mechanisms and selected properties of floors are presented. It was emphasized that the inevitability of the degradation processes progressing over time, reducing the safety of use and durability, necessitates taking preventive measures based on diagnostic procedures.

REFERENCES

1. EN 1990:2004 Basics of structure design, 2004
2. De Schutter, G. (2012). Damage to concrete structures. Boca Raton, FL: CRC Press.
3. Kmiecik, P., & Kamiński, M. (2011). Modelling of reinforced concrete structures and composite structures with concrete strength degradation taken into consideration. Archives of civil and mechanical engineering, 11(3), 623–636.
4. Glasser, F.P., Marchand, J., & Samson, E. (2008). Durability of concrete—Degradation phenomena involving detrimental chemical reactions. Cement and concrete research, 38(2), 226–246.
5. Sarja, A., & Vesikari, E. (Eds.). (2004). Durability design of concrete structures. Boca Raton, FL: CRC Press.
6. Ballarini, R., Gencturk, B., Jain, A., Aryan, H., Xi, Y., Abdelrahman, M., & Spencer, B.W. (2020). Multiple degradation mechanisms in reinforced concrete structures, modeling and risk analysis (No. INL/EXT-20-57095-Rev000). Idaho Falls, ID: Idaho National Lab.(INL).
7. Czarnecki, L., Łukowski, P., & Garbacz, A. (2017). Repair and protection of concrete structures - commentary to PN-EN 1504 (in Polish). Warsaw: Polish Scientific Publishers PWN.

8. Czarnecki, L., & Emmons, P. (2002). Repair and protection of concrete structures (in Polish). Kraków: Polski Cement.

9. Golewski, G.L. (2016). Reinforced concrete structures loaded dynamically. Lublin: Lublin University of Technology Publishing House.

10. Poonguzhali, A., Shaikh, H., Dayal, R.K., & Khatak, H.S. (2008). A review on degradation mechanism and life estimation of civil structures. Corrosion reviews, 26(4), 215–294.

11. Bertron, A. (2014). Understanding interactions between cementitious materials and microorganisms: A key to sustainable and safe concrete structures in various contexts. Materials and structures, 47(11), 1787–1806.

12. Bertolini, L. (2008). Steel corrosion and service life of reinforced concrete structures. Structure and infrastructure engineering, 4(2), 123–137.

13. Runkiewicz, L. (2016). Strengthening reinforced concrete structures (in Polish). Poradnik. Warsaw: Building Research Institute.

14. Runkiewicz, L., & Kowalewski, J. (1999). Principles of safety assessment of reinforced concrete structures (in Polish). Warsaw: Building Research Institute.

3 Introductory Information to the Diagnostics of Floors

3.1 TYPES OF TECHNICAL STUDIES

As was stated at the end of Chapter 2, technical studies are the result of technical inspections of floors, both periodic (cyclical) and those carried out in necessary situations [1–4]. Such studies, depending on the purpose they are to serve, differ in terms of the name of the scope, detail, and inquisitiveness. Figure 3.1 lists the names of technical studies that are customary in construction practice – they have not been defined in the regulations.

It is therefore customary to assume that:

- technical evaluation, which is the simplest of the studies listed in Figure 3.1, concerns the assessment of specific events, phenomena or processes, and includes an assessment of the threats and the condition of the floor. This study usually describes and analyses the expected (designed) technical condition, and is based on design principles, and not on the results of diagnostic tests,
- technical opinion concerns the assessment of specific design solutions, events or phenomena occurring in the process of construction implementation or use. It may include the judgement of material solutions and financial outlays. Technical opinion expresses an opinion or belief about something, and is not supported by a research process (diagnostics), therefore often being subjective,
- technical judgement is a study that is broader in terms of scope and detail than the technical assessment and opinion. It concerns the assessment of technical solutions, phenomena, and events occurring in the process of the design, implementation, and use of floors. It may include the assessment of individual structural elements, as well as the assessment of technological and material solutions. These assessments are usually supported by the results of diagnostic procedures. In the case of unfavourable events or phenomena, the decision specifies the reasons for their occurrence and formulates the final assessment,
- technical expertise is a study of the broadest scope (detailed and inquisitive) and is based primarily on the results of diagnostic procedures and the research included in these procedures. It documents and assesses the phenomena, events, and processes occurring during the construction or use of the floor. The aim of the expert's opinion is to determine the current

DOI: 10.1201/9781003288374-3

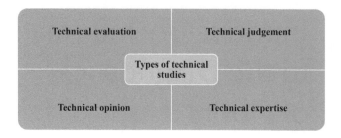

FIGURE 3.1 Types of technical studies.

technical condition of the assessed floor after the occurrence of circumstances, the effect of which is, for example, the formation of displacements, cracks, or excessive deflections. Its task is also to indicate the causes of irregularities, and to indicate effective methods of their removal, etc. The expertise also includes the identification and location of any damage and defects of the assessed floor, the testing of selected features embedded in the floor, and mechanical and structural tests of these materials, etc. For this purpose, diagnostic procedures, which use non-destructive and destructive methods, are usually tailored to specific needs. It is worth noting that the technical expertise may also include a computational verifying static analysis of either individual elements or the entire floor system. The conclusions resulting from the technical expertise should constitute the basis for determining the further course of action concerning the assessed floor.

Typically, the layout of the technical expertise is illustrated in Figure 3.2, and is included in several sections (marked for the purpose of this chapter) with Arabic numerals in parentheses. It is worth briefly commenting on this arrangement (following work [5]).

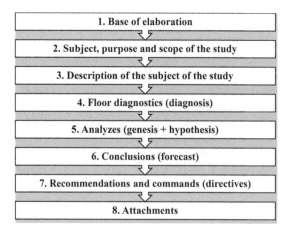

FIGURE 3.2 Usual layout of the content of technical expertise.

With regard to the basis of the study (1), the following should be provided: the formal basis for the understanding of the data on the order and the concluded contract, dates of on-site visits and visual tests, materials used for the study, detailing the data received from the client or obtained from another source, the literature used, standards, recipes, and recommendations.

The subject, purpose, and scope of the studies (2) should be specified clearly and precisely, because deficiencies in this regard may be used in the event of civil law disputes.

The description of the subject of the study (3) should be concise, and supplemented, if necessary, with drawings or sketches and photographs that include the necessary dimensions. The foundation method, soil and water conditions, and possible environmental impacts should also be specified.

The scope of diagnostics (4) should be selected according to the purpose of the expertise. In the case of floors, it may be wide and include, among others: the analysis of technical documentation, the assessment of the technical condition of the structure or its fragments, various tests of structures or their fragments, geotechnical tests of the subsoil, geodetic and inventory tests of the structure, specialist tests of concrete and reinforcing steel in situ and in the laboratory, environmental tests, and tests identifying defects and damages. In addition, test loads may also be included. Bearing in mind the numerous cases from construction practice that show that a structure works in a way not as it was designed, but as it was made, it is also reasonable to identify static patterns and loads on the structure as part of the performed expertise performed [5]. For the purposes of diagnostics, modern non-destructive and semi-destructive testing methods should first be used. These diagnostic activities, which are not a closed set, are the key to explaining the causes of the situations or phenomena that have occurred [6].

Analyses of occurring phenomena, for example, in the form of existing and identified structural damages and defects, as well as the determination of the reasons for their occurrence will be included in Chapter (5). It is based on the results of measurements and diagnostic tests of floors that are also helpful in carrying out the necessary checking calculations, including static, strength, structural reliability, maintenance, and serviceability limit states. If the comprehensiveness of calculations, including numerical ones, is significant, transferring them to the appendix should be considered (8), with the results being left in this chapter in the form of, for example, tabular summaries. As a result of calculations and analyses, the difference between the existing and the desired state is determined, with the conditions of emergence and the meaning of the phenomena (genesis) on which the basis of the hypotheses that explain the phenomena being formulated.

Chapters (6) and (7) of the content of the technical expertise arranged in this way are, in order, conclusions (forecasts) formulated on the basis of the results of diagnostic activities, calculations, and analyses obtained in (4) and (5), as well as recommendations and orders (directives). These sections of expert opinions usually formulate the course of further proceedings regarding a floor, concerning, among others: improvement of the existing technical condition, meeting the basic requirements, concepts and guidelines for repairs, reinforcements, demolition, etc.

In the appendices (8) it is worth including, inter alia, photographic documentation, drawing documentation with an inventory of defects and damage (with the places of the conducted excavations marked), detailed results of diagnostic tests, extensive calculations, etc. Placing these "materials" in accordance with the presented diagram (Figure 3.2) has a positive effect on the readability of the expert opinion.

3.2   TYPES OF DIAGNOSTICS

The term diagnostics comes from the Greek language (diagnosis) and in general terms refers to the assessment of the technical condition of a structure, as well as the prognosis and reasons for its development or change [6].

Floor diagnostics is carried out, mentioned in Section 3.1, primarily during periodic (periodic) technical inspections, and has a significant share in the efforts to take care of the safety of use and the maintaining of the projected durability of a structure. In addition, it is carried out in necessary situations when there are unintentional violations of the structure caused by a change in operating conditions, overload, fire, etc.

In construction practice, there are basically three types of diagnostics of floors: (1) periodic, (2) ad hoc, and (3) target. In addition, there are one-stage and two-stage diagnostics. Therefore:

- Periodic diagnostics is carried out during periodic technical inspections on dates most often resulting from applicable law or arrangements made by the user,
- Ad hoc diagnostics is carried out in emergency situations, which include damage to the structure or its individual elements, and the need for execution is indicated by the user or by technical supervision or state construction supervision,
- Target diagnostics is usually carried out when there is a need to assess the possibilities and conditions of the planned reconstruction, change of use, modernization, etc.

When it comes to single-stage diagnostics, it should be understood as normal, performed according to schedule, and uncaused, for example, due to the poor condition of the structure, the need to take quick actions in connection with the planned modernization or change in the way of use, or the need for temporary reinforcement. When quick decisions are needed, two-stage diagnostics are performed [6–10].

3.3   SELECTION OF DIAGNOSTIC METHODS

In the diagnosis of concrete structures, a key role is played by the test methods, the selection of which should be adapted to the purpose and technical condition of the diagnosed floor. The methods used to diagnose the floor should make it possible to obtain reliable test results with the required accuracy that are useful for characterizing the condition of the structure [11]. Among the numerous research methods, non-destructive and semi-destructive methods are especially recommended, the main advantage of which is the lack or very little interference in the tested floors. These non-invasive or minimally invasive methods, apart from not interfering with the examined floor, can be used for examinations in any number of places and at any time [12]. These methods are covered in Chapters 4–6.

3.4   GENERAL PRINCIPLES OF THE DIAGNOSTIC PROCEDURE

The results of diagnostic procedures constitute the basis for the prepared technical studies of an expert opinion. The general principles that should be taken into account have been listed below (according to the authors of works [5, 6]).

Existing concrete floors should be assessed according to current standards. Earlier standards, which were in force at the time of designing these structures, can only be the so-called Informative "background". The levels of safety and reliability of the floor should not be lower than the levels required for designing new floors.

Depending on the specific needs, diagnostic procedures can be applied to either the entire floor or only to some of its elements, for example, damaged parts.

In tests of floors or their components, the actual condition of the floor and the working conditions should be taken into account.

The type of diagnostics of floors should be adapted to the purpose that the particular diagnosis is to serve.

The selection of diagnostic methods and techniques should be adapted to the conditions and technical condition of the diagnosed floors. In a situation where the tests performed with non-destructive and semi-destructive methods turn out to be insufficient, it is worth supporting them with test loads.

The applied non-destructive and semi-destructive methods should ensure that the obtained results have a required accuracy and are numerous enough to enable the statistical evaluation required by the standard regulations.

Tests of floors, both in operation and out of use, should be preceded by an analysis of the available design documentation and other documents, which include, inter alia, a construction log, facility book, past opinions, and judgements and technical expertise regarding the floor. The source of information may also be interviews with construction contractors, its users and managers, state construction supervisors, etc.

During the tests, attention should be paid to factors that may not directly affect the limit states of the bearing capacity or serviceability of the assessed floor, but which indirectly, in the long term, may significantly affect its safety and reliability. Examples of such factors are, among others, an aggressive environment, a lack of anti-dampness protection, and vibrations transmitted to the structure from the environment.

For the safety and reliability of floors, it is very important to evaluate them based on actual data, including: the dimensions of elements: the properties and characteristics of the concrete, including its strength and homogeneity; the properties and features of the reinforcing steel, including its distribution: the concrete cover; corrosion; loads and their nature; models (diagrams) of work; deformation; and the identification and location of damage and defects.

3.5   CONCLUSIONS

This chapter gathers and presents introductory information to the diagnosis of concrete floors. It was emphasized that floor diagnostics is usually carried out during periodic technical inspections and that the most detailed and inquisitive result is a technical expertise based on the results of diagnostic procedures and tests. The types of diagnostics distinguished in construction practice were discussed and the problem of the proper selection of appropriate research methods was signalled. In addition, this chapter provides the principles of diagnostic procedures.

REFERENCES

1. Runkiewicz, L. (2016). Strengthening reinforced concrete structures. Guide (in Polish). Warsaw: Building Research Institute.
2. Abdou, A., Haggag, M., & Al Khatib, O. (2016). Use of building defect diagnosis in construction litigation: Case study of a residential building. Journal of legal affairs and dispute resolution in engineering and construction, 8(1), C4515007.
3. Zoidis, N., Tatsis, E., Vlachopoulos, C., Gotzamanis, A., Clausen, J.S., Aggelis, D.G., & Matikas, T.E. (2013). Inspection, evaluation and repair monitoring of cracked concrete floor using NDT methods. Construction and building materials, 48, 1302–1308.
4. Garcia, J., & De Brito, J. (2008). Inspection and diagnosis of epoxy resin industrial floor coatings. Journal of materials in civil engineering, 20(2), 128–136.
5. Bortolini, R., & Forcada, N. (2018). Building inspection system for evaluating the technical performance of existing buildings. Journal of performance of constructed facilities, 32(5), 04018073.
6. Czapliński, K. (2012). The method and form of developing construction expertise (in Polish). Wrocław: Lower Silesian Educational Publisher.
7. Drobiec, Ł., Jasiński, R., & Piekarczyk, A. (2010). Diagnostics of reinforced concrete structures (in Polish). Warsaw: Polish Scientific Publishers PWN.
8. François, R., Laurens, S., & Deby, F. (2018). Corrosion and its consequences for reinforced concrete structures. London and Kidlington, UK: ISTE Press – Elsevier.
9. Delgado, J.M. (Ed.). (2021). Durability of concrete structures (Vol. 16). Cham, Switzerland: Springer Nature.
10. Bungey, J.H., Millard, S.G., & Grantham, M.G. (2006). Testing of concrete in structures. Boca Raton, FL: CRC Press.
11. Runkiewicz, L. (1999). Diagnostics and strengthening of reinforced concrete structures (in Polish). Kielce: Publishing House of the Kielce University of Technology.
12. Panasyuk, V.V., Marukha, V.I., & Sylovanyuk, V.P. (2014). Methods and devices for technical diagnostics of long-term concrete structures. In Injection technologies for the repair of damaged concrete structures (pp. 185–206). Dordrecht: Springer.

4 The Classification of Methods for Floor Diagnostics

4.1 NON-DESTRUCTIVE METHODS

Various non-destructive methods and techniques are used in concrete structures diagnostics. Attempts to classify them, as well as their descriptions, can be found, among others, in the works [1–11]. Based on the analysis of these works and the authors' own experiences, Figure 4.1 presents the basic groups of the non-destructive methods that are useful in floor diagnostics, namely: physical methods, chemical methods, and biological methods. The group of physical methods uses physical phenomena, chemical methods use chemical processes, with biological methods being related to biological processes.

The group of physical methods is the most numerous and the most commonly used in floor diagnostics, and it has therefore been treated with more detail in relation to the other groups. It can be divided into the eight subgroups listed in Figure 4.1. In each of these subgroups, a number of methods are used, which are detailed in Figure 4.2.

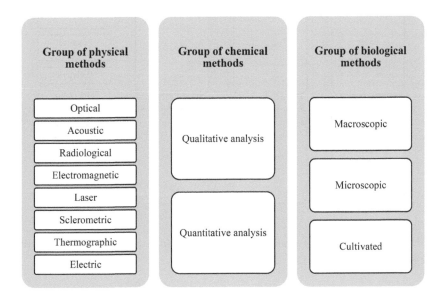

FIGURE 4.1 Basic groups of non-destructive methods useful in floor diagnostics.

DOI: 10.1201/9781003288374-4

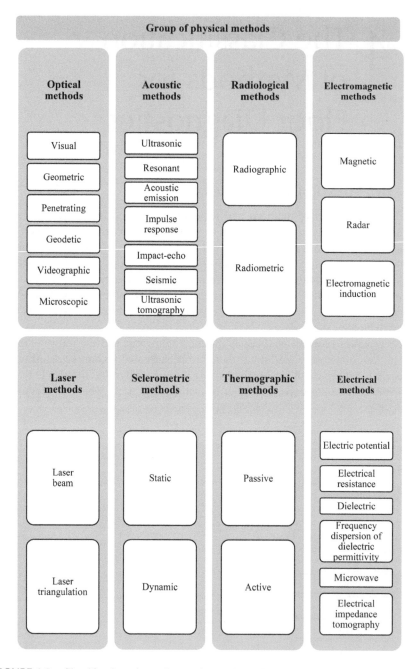

FIGURE 4.2 Classification of non-destructive methods from the group of physical methods useful in floor diagnostics.

4.2 USABILITY AREAS OF NON-DESTRUCTIVE METHODS

For the groups of non-destructive, physical, chemical, and biological methods distinguished in Figure 4.1, the areas of usefulness in floors diagnostics were determined in work [1]. The methods that are classified as physical groups are shown in Table 4.1. Table 4.2 shows the methods classified as chemical and biological, and specifies the areas of usefulness in research with regards to geometry, material properties, and the detection and identification of damage, along with information on which methods are considered basic in the sense of universal application, and which are complementary.

In floor diagnostics, the visual assessment of the surface of these structures plays an important role. Figure 4.3 breaks down the non-destructive methods that are useful for the visual assessment of the surface of into direct and indirect.

As floor diagnostics very often assess the strength of concrete and the homogeneity of the distribution of strength, Figure 4.4 lists the non-destructive methods that are useful for this purpose. These are mainly static and dynamic sclerometric methods, and also some of the acoustic methods, in particular the ultrasonic and resonance methods.

Methods that are useful for assessing the elements accessible from one side and determining the location of defects and damage invisible on the surface are also very often used in floor diagnostics. Therefore, in Figure 4.5, and with reference to Table 4.1, these methods are summarized. The most suitable for these purposes are acoustic methods, such as impulse response, impact-echo, seismic, ultrasonic and acoustic emission; radiological methods; and electromagnetic methods such as radar methods.

Floors made of concrete are predominantly reinforced with reinforcing steel. Therefore, Figure 4.6 shows the non-destructive methods used to locate the reinforcement and assess its corrosion. For these purposes, radiological and electrical methods, as well as electromagnetic methods, are most useful.

When diagnosing floors, it is also very often necessary to know the moisture and temperature of the concrete, and in some cases also the distribution of these parameters on the surface, with the methods useful for this purpose being given in Figure 4.7. Among the electrical methods, the most useful are dielectric, frequency dispersion, dielectric permittivity, microwave, and electrical impedance tomography. Of the optical methods, videography is useful. Among the thermographic methods, passive and active thermography are useful, and among the radiological methods, radiographic and radiometric are useful.

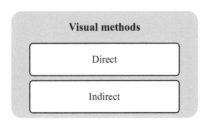

FIGURE 4.3 Non-destructive methods useful for the visual assessment of the surface of floors.

TABLE 4.1

Areas of Usefulness of Non-Destructive Methods from the Group of Physical Methods in Floor Diagnostics (Own Study Based on [1])

	Usability Areas						
	For the Study of Geometry			For the Study of Material Properties			
Name of the Method	Spatial Configuration	Dimensions of the Elements	Reinforcement Identification	Type of Material	Strength/ Homogeneity	Modulus of Deformation	Porosity/ Water Absorption
Optical Methods							
Visual	■	□	-	■	□	-	-
Geometric	■	■	□	-	-	-	-
Penetrating	-	-	-	□	■	-	□
Geodetic	■	■	-	-	-	-	-
Videographic	-	-	-	-	-	-	-
Microscopic	-	-	-	-	-	-	-
Acoustic Methods							
Ultrasonic	-	■	-	-	-	-	■
Resonant	-	-	-	-	-	-	-
Acoustic emission	-	-	-	-	-	-	-
Impulse responses	-	■	-	-	-	-	■
Impact-echo	-	■	-	-	-	-	■
Seismic	-	-	-	-	-	-	-
Ultrasound tomography	-	-	□	-	-	-	-
Radiological Methods							
Radiographic	-	-	■	□	□	-	□
Radiometric	-	-	□	□	□	-	□
Electromagnetic Methods							
Magnetic	-	-	■	-	-	-	-
Radar	□	□	■	-	-	-	-
Electromagnetic induction	-	-	■	-	-	-	-
Laser Methods							
Laser beam	■	■	-	-	-	-	-
Laser triangulation	■	■	-	-	-	-	-
Sclerometric Methods							
Static	-	-	-	-	■	□	-
Dynamic	-	-	-	-	■	□	-
Thermographic Methods							
Passive	-	-	□	-	-	-	-
Active	-	-	□	-	-	-	-
Electrical Methods							
Electric potential	-	-	-	-	-	-	-
Electrical resistance	-	-	-	-	-	-	-
Dielectric	-	-	-	-	-	-	-
Frequency dispersion of dielectric permittivity	-	-	-	-	-	-	-
Microwave	-	-	-	-	-	-	-
Electrical impedance tomography	-	-	-	-	-	-	-

Designation: ■ – Basic method; □ – Complementary method; - Not applicable

(*continued*)

TABLE 4.1 (*Continued*)
Areas of Usefulness of Non-Destructive Methods from the Group of Physical Methods in Floor Diagnostics (Own Study Based on [1])

| | | | | Usability Areas | | | | |
| | | | | For Damage Detection | | | | |
Frost Resistance	Temperature and Humidity	Chemical Composition	Deformations	Destruction of the Material	Material Losses	Loss of Material Continuity	Impurities	Changes in Location
-	□	-	■	□	■	■	■	■
-	-	-	■	□	■	■	■	■
-	-	-	-	□	□	-	-	-
-	-	-	■	-	□	-	-	■
-	■	-	-	-	-	-	-	-
-	-	-	-	■	-	-	■	-
■	-	-	-	□	■	■	-	-
-	-	-	-	-	-	■	-	-
■	-	-	-	□	■	■	-	-
■	-	-	-	□	■	■	-	-
-	-	-	-	□	-	-	-	-
-	-	-	-	□	■	■	-	-
-	-	-	-	□	■	■	-	-
-	-	-	-	□	■	■	-	-
-	-	-	-	-	-	-	-	-
-	-	-	-	□	■	■	-	-
-	-	-	-	-	-	-	-	-
-	-	-	■	-	-	-	-	■
-	-	-	■	-	-	-	-	■
-	-	-	-	□	-	-	-	-
-	-	-	-	□	-	-	-	-
-	□	-	-	□	□	□	-	-
-	□	-	-	□	□	□	-	-
-	□	-	-	■	■	-	-	-
-	■	-	-	■	■	-	-	-
-	■	-	-	-	-	-	-	-
-	■	-	-	-	-	-	-	-
-	■	-	-	-	-	-	-	-
-	■	-	-	-	-	-	-	-

TABLE 4.2

Areas of Usefulness of Non-Destructive Methods from the Group of Chemical and Biological Methods in Floor Diagnostics (Own Study Based on [1])

	Usability Areas						
	For the Study of Geometry			For the Study of Material Properties			
Name of the Method	Spatial Configuration	Dimensions of the Elements	Reinforcement Identification	Type of Material	Strength/ Homogeneity	Modulus of Deformation	Porosity/ Water Absorption
Group of Chemical Methods							
Qualitative analysis methods	-	-	-	□	-	-	-
Quantitative analysis methods	-	-	-	■	-	-	-
Group of Biological Methods							
Macroscopic methods	-	-	-	-	-	-	-
Microscopic methods	-	-	-	-	-	-	-
Cultivated methods	-	-	-	-	-	-	-

Designation: ■ – Basic method; □ – Complementary method; - Not applicable

(*continued*)

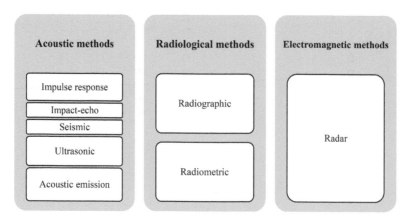

FIGURE 4.4 Non-destructive methods useful for the assessment of concrete strength and the homogeneity of its distribution in the floor.

FIGURE 4.5 Non-destructive methods useful for assessing the thickness of floors and the location of defects and damage.

TABLE 4.2 (*Continued*)
Areas of Usefulness of Non-Destructive Methods from the Group of Chemical and Biological Methods in Floor Diagnostics (Own Study Based on [1])

				Usability Areas				
				For Damage Detection				
Frost Resistance	Temperature and Humidity	Chemical Composition	Deformations	Destruction of the Material	Material Losses	Loss of Material Continuity	Impurities	Changes in Location
-	-	□	-	□	-	-	□	-
-	-	■	-	■	-	-	■	-
-	-	-	-	■	-	-	■	-
-	-	-	-	■	-	-	■	-
-	-	-	-	□	-	-	□	-

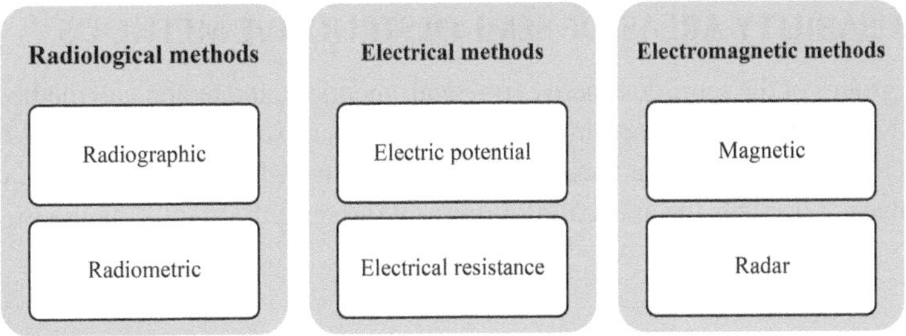

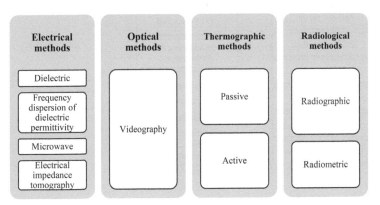

FIGURE 4.6 Non-destructive methods useful for the location of reinforcement in a floor and the assessment of its corrosion.

FIGURE 4.7 Non-destructive methods useful for assessing the moisture and temperature of concrete and the distribution of these parameters on the surface.

FIGURE 4.8   Non-destructive methods useful for the assessment of the morphology of floors.

When diagnosing floors, it is necessary in some cases to know the morphology of the surface. Figure 4.8 lists the methods useful for this purpose.

4.3   SEMI-DESTRUCTIVE METHODS

Similarly to non-destructive methods, semi-destructive methods are also very useful in many research situations in floor diagnostics. They are used primarily in the field (in situ research), but also in laboratory conditions. Their use does not significantly affect the structure of the tested floors, which is usually only point-like. Figure 4.9, according to [11], is a proposal of a general classification of the basic methods and techniques of non-destructive failure in floor diagnostics in the field (in situ). As in the case of non-destructive methods, physical, chemical, and biological methods have been distinguished in this classification.

4.4   USABILITY AREAS OF SEMI-DESTRUCTIVE METHODS

For the groups of the semi-destructive (physical, chemical, and biological) methods listed in Figure 4.9, the areas of usefulness in the diagnosis of selected properties of concrete were determined. They were based, similarly to the non-destructive methods, on work [11]. Table 4.3 presents the areas of usefulness of the semi-destructive methods.

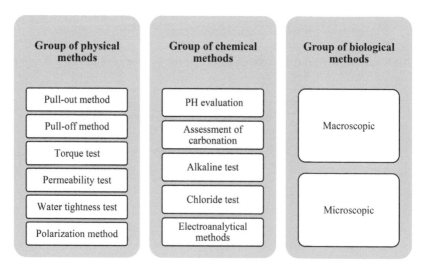

FIGURE 4.9   Classification of semi-destructive methods useful in floor diagnostics in field conditions.

TABLE 4.3

Areas of Usefulness of Semi-Destructive Methods in the Diagnosis of Selected Properties of Floors under Field Conditions (Own Elaboration Based on [11])

Name of the Method	Usability Areas				
	In the Study of the Structure of the Material	In the Study of the Mechanical Parameters of the Material	In the Study of Material Discontinuities	In Water and Gas Permeability Tests	In the Study of the Chemical Composition
Group of Physical Methods					
Pull-out method	-	■	-	-	-
Pull-off method	-	■	-	-	-
Torque test	-	■	-	-	-
Permeability test	-	■	-	-	-
Water tightness test	□	-	-	■	-
Polarization method	-	-	-	-	■
Group of Chemical Methods					
PH evaluation	-	-	-	-	■
Assessment of carbonation	-	-	-	-	■
Alkaline test	-	-	-	-	■
Chloride test	-	-	-	-	■
Electroanalytical methods	-	-	-	-	■
Group of Biological Methods					
Macroscopic	□	-	□	-	-
Microscopic	□	-	□	-	-

Designation: ■ – Basic method; □ – Complementary method; - Not applicable

A proposal for a general classification of basic semi-destructive methods and techniques useful in floor diagnostics under laboratory conditions, according to [11], is shown in Figure 4.10. Physical, chemical, and biological methods were also distinguished. They are used in the laboratory testing of samples, including, for example, for core samples taken from the structure. Table 4.4 provides information on the suitability of these methods in the diagnosis of selected properties of floors.

4.5 CONCLUSIONS

This chapter proposes classifications of non-destructive and semi-destructive methods. These classifications distinguish groups of these methods with the division into physical, chemical, and biological. Much attention was paid to the group of physical

TABLE 4.4

Usefulness of Semi-Destructive Methods in Floor Diagnostics under Laboratory Conditions (Own Elaboration Based on [11])

Name of the Method	Usability Areas				
	In the Study of the Structure of the Material	In the Study of the Mechanical Parameters of the Material	In the Study of Material Discontinuities	In Water and Gas Permeability Tests	In the Study of the Chemical Composition
Group of Physical Methods					
Strength tests	-	■	-	-	-
Research on elastic parameters	-	■	-	-	-
Water absorption tests	□	-	-	■	-
Porosity tests	■	-	-	■	-
Frost resistance tests	-	■	-	-	-
Abrasion tests	□	■	-	-	-
Structure tomographic research	■	□	□	-	-
Group of Chemical Methods					
Electroanalytical	-	-	-	-	■
Spectral analysis	-	-	-	-	■
Chromatographic	-	-	-	-	■
Group of Biological Methods					
Microbiological analysis	□	-	□	-	-
Cultivated	□	-	□	-	-

Designation: ■ – Basic method; □ – Complementary method; - Not applicable

methods most commonly used in the diagnosis of concrete floors, distinguishing numerous subgroups therein and assigning the named methods to each subgroup. In order to facilitate the selection of the appropriate method for a specific research need (situation), areas of suitability in the diagnosis of concrete floors have been assigned in this chapter.

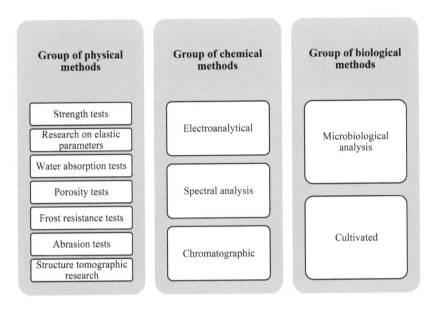

FIGURE 4.10   Classification of semi-destructive methods useful in floor diagnostics under laboratory conditions.

REFERENCES

1. Bień, J. (2010). Defects and diagnostics of bridge structures (in Polish). Warsaw: Wydawnictwo Komunikacji i Łączności.
2. Malhotra, V.M., & Carino, N.J. (2003). Handbook on nondestructive testing of concrete. Boca Raton, FL: CRC Press.
3. Bungey, J.H., Millard, S.G., & Grantham, M.G. (2006). Testing of concrete in structures. Boca Raton, FL: CRC Press.
4. Blitz, J., & Simpson, G. (1995). Ultrasonic methods of non-destructive testing (Vol. 2). London, UK: Chapman & Hall.
5. Maierhofer, C., Arndt, R., Röllig, M., Rieck, C., Walther, A., Scheel, H., & Hillemeier, B. (2006). Application of impulse-thermography for non-destructive assessment of concrete structures. Cement and concrete composites, 28(4), 393–401.
6. Pucinotti, R. (2015). Reinforced concrete structure: Non destructive in situ strength assessment of concrete. Construction and building materials, 75, 331–341.
7. Maierhofer, C., Reinhardt, H.W., & Dobmann, G. (Eds.). (2010). Non-destructive evaluation of reinforced concrete structures: non-destructive testing methods. Sawston, UK: Woodhead Publishing.
8. Dobmann, G., Maierhofer, C., & Reinhardt, H.W. (Eds.). (2010). Non-destructive evaluation of reinforced concrete structures: Deterioration processes and standard test methods. Boca Raton, FL: CRC Press.
9. Hoła, J., & Schabowicz, K. (2010). State-of-the-art non-destructive methods for diagnostic testing of building structures–anticipated development trends. Archives of civil and mechanical engineering, 10(3), 5–18.
10. Drobiec, Ł., Jasiński, R., & Piekarczyk, A. (2014). Diagnostics of concrete structures (in Polish). Warsaw, Poland: Polish Scientific Publishers PWN.
11. Hoła, J., Bień, J., Sadowski, Ł., & Schabowicz, K. (2015). Non – destructive and semi - destructive diagnostics of concrete structures in assessment of their durability. Bulletin of the Polish academy of science. Technical sciences, 63(1), 87–96.

5 The Overview of Methods Useful in Floor Diagnostics

5.1 NON-DESTRUCTIVE METHODS USEFUL FOR VISUAL ASSESSMENT

Non-destructive methods useful for the visual assessment of concrete floors include measurement techniques that use simple instruments such as magnifying glasses, as well as optical test equipment such as endoscopes, borescopes, and videoscopes, which enable the examination of surfaces and places inaccessible to the eye [1–4]. These methods are divided into direct and indirect methods, as shown in Table 5.1.

TABLE 5.1

Synthetic Description of the Non-Destructive Methods Useful for the Visual Assessment of Concrete Floors

Name and Description of the Method	View of the Apparatus Used	Recorded Parameters, Exemplary Test Results
The visual methods rely on the visual assessment of concrete floors. The direct method is used to examine the surface directly accessible to the eye, and the examination does not require the use of special instruments – at most a magnifying glass. The indirect method (optical tests) is used to examine surfaces and places inaccessible directly to the eye, and the testing uses devices equipped with special cameras, for example, endoscopes, borescopes, videoscopes.	 	s – scratch width, L – scratch length. The image is visible with the use of a magnifying glass: 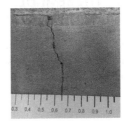 Image captured by an endoscope camera: 

DOI: 10.1201/9781003288374-5

29

5.2   NON-DESTRUCTIVE METHODS USEFUL FOR ASSESSING THE STRENGTH OF CONCRETE AND ITS DISTRIBUTION IN A FLOOR

As mentioned in Section 4.1 in Chapter 4, sclerometric methods and some of the acoustic methods, in particular the ultrasonic method, are especially useful for the non-destructive evaluation of the strength of concrete and its distribution in a floor. A synthetic description of the dynamic sclerometric method and the ultrasonic method is presented in Table 5.2.

5.3   NON-DESTRUCTIVE METHODS USEFUL FOR ASSESSING THE THICKNESS OF FLOORS AND THE LOCATION OF DEFECTS

Of the non-destructive acoustic test methods listed in Chapter 4, the most useful for assessing the thickness of floors and the location of defects are the impulse response method, the impact-echo method and the ultrasonic tomography method. These methods are discussed below in Table 5.3.

5.4   NONDESTRUCTIVE METHODS USEFUL FOR THE LOCATION OF REINFORCEMENT

A synthetic description of selected methods useful for the location of reinforcement is presented in Table 5.4. Among these methods, the magnetic and radar methods are discussed.

5.5   NON-DESTRUCTIVE METHODS USEFUL FOR THE ASSESSMENT OF THE MORPHOLOGY OF A CONCRETE FLOOR SURFACE (SEE TABLE 5.5)

5.6   THE SEMI-DESTRUCTIVE METHOD USEFUL IN IN SITU RESEARCH

The methods useful in field conditions, in in situ tests, include the pull-off method. This method allows the subsurface tensile strength of concrete floor or the interlayer tensile adhesion to be assessed, and involves measuring the detachment forces of metal discs glued to the concrete floor surface. The method is described in [26, 27], and is standardized in [26, 28]. A synthetic description of the method is presented in Table 5.6.

5.7   NON-DESTRUCTIVE METHODS USED IN THE LABORATORY

The non-destructive methods useful in laboratory conditions include the method of assessing the structure of concrete samples taken from a floor. A synthetic description of selected non-destructive methods useful for this purpose is presented in Table 5.7.

TABLE 5.2

Synthetic Description of Selected Non-Destructive Methods Useful for the Assessment of Concrete Strength and Its Distribution in the Floor

Name and Description of the Method	View of the Apparatus Used	Recorded Parameters, Exemplary Test Results
The sclerometric (dynamic) method involves the measuring, by assessing the energy change of a metal mandrel, the rebound of the mandrel from the tested surface. In this method, it is required to define a parameter in the form of the so-called rebound number R and then to correlate it with the compressive strength f_c by selecting an appropriate correlation or hypothetical relationship f_c-R. In order to select such a relationship, it is required to collect a certain number of concrete samples from the floor for strength and sclerometric tests. The correct application of this method, ensuring that reliable results are obtained, requires the absolute application of the principles of conducting research (collected in works [5–7]). Work [5] provides a detailed procedure for in situ concrete testing. In addition, standard [8] also applies.	 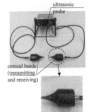	R – rebound number. Sample table with the results of work [9]: Examples of the correlation relationships used in [9]: 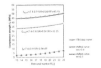
The ultrasonic (classical) method involves the generating and transmitting of longitudinal ultrasonic waves, followed by their registration after passing through the tested concrete floor. The apparatus consists of an ultrasonic probe and ultrasonic heads (transmitting and receiving) of various shapes (cylindrical, conical, exponential) and different frequencies (from 25 to 1000 kHz). Due to the way the probes are applied, three basic methods of testing should be distinguished: through, surface and corner. To evaluate strength, it is required to determine the velocity of the ultrasonic longitudinal waves propagating in the tested concrete by measuring their transition time t from the transmitting head to the receiving head (path l), and then correlating them with the compressive strength obtained on the basis of strength tests of samples taken from the floor. The results of measurements of the strength of concrete in the floor should be verified by calibrating the method, which involves selecting an appropriate correlation or hypothetical relationship. Correct application of these methods requires absolute observance of the principles of conducting research (collected in [5], as well as in [10]), which provides a detailed procedure for ultrasonic testing of concrete in situ. Moreover, standards [11, 12] are in force. When cylinder heads are used, it is necessary to use a coupling agent between the cylinder heads and the concrete surface.	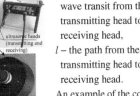	t – time of the ultrasonic wave transit from the transmitting head to the receiving head, l – the path from the transmitting head to the receiving head. An example of the course of the concrete compressive strength f_c as a function of the thickness h of the floor correlated with the ultrasonic wave velocity c_L obtained from the tests obtained by the ultrasonic method with the use of conical heads: 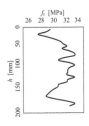

TABLE 5.3

Synthetic Description of Selected Methods Useful for Assessing the Thickness of Floors and the Location of Defects

Name and Description of the Method	View of the Apparatus Used	Recorded Parameters, Exemplary Test Results
The **impulse response method** involves inducing an ultrasonic elastic wave in the tested element, which is calibrated with a hammer that has a built-in rubber tip. The frequency of the induced wave is in the range from 1 to 800 Hz, while the excitation range around the test point is up to about 1000 mm. The apparatus consists of a hammer with a rubber tip and a geophone, and the received signal is amplified by an amplifier and recorded on a laptop. A quick search of large concrete floors, and the rough identification and surface location of defects in the form of delamination, cracks, inclusions and air voids is in practice carried out with this method at test sites. A grid of measuring points with a spacing of 500 to 1000 mm is formed.		N_{av} – average mobility, K_d – dynamic stiffness, M_p/N – mobility slope, $N_{av} \cdot M_p/N$ – average mobility times mobility slope, v – voids index. An exemplary map of the average mobility N_{av} prepared for a fragment of a concrete floor: 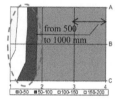
The **impact-echo method** is based on the excitation of an elastic wave of low frequency from 1 to 60 kHz in the tested element by hitting its surface with an inductor in the form of a steel ball. The equipment includes a set of steel ball inductors of various diameters, a measuring (receiving) head and a laptop with specialized software. Specialized software allows a graphic image of an elastic wave propagating in the tested floor to be recorded (in the amplitude-time system), and for this image to be converted into an amplitude-frequency spectrum using the fast Fourier transform. The impact-echo method is suitable for testing concrete floors and assessing: thickness when there is only one-sided access, the depth of scratches visible on the surface, adhesion between concrete layers, delamination, cracks, foreign inclusions, air voids, the efficiency of cable channel injection in post-reinforced concrete elements, etc. [13, 14].	 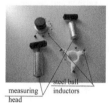	A – amplitude, f_D – frequency of reflection of the ultrasonic wave from the defect, f_T – frequency of reflection of the ultrasonic wave from the bottom of the concrete floor. An example of an amplitude-frequency diagram in the case of reflection of an ultrasonic wave from the bottom of a concrete floor and from defects: 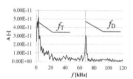

(continued)

TABLE 5.3 (*Continued*)

Synthetic Description of Selected Methods Useful for Assessing the Thickness of Floors and the Location of Defects

Name and Description of the Method	View of the Apparatus Used	Recorded Parameters, Exemplary Test Results
Ultrasonic tomography method involves the excitation of elastic waves in the tested floor. The inductor is a multi-head antenna with several dozen independent ultrasonic heads built into it, which is also used to receive and process ultrasonic signals. The heads generate ultrasonic pulses with a frequency of 50 kHz. This method does not require the use of a coupling agent at the joint of the heads with the concrete surface. The ultrasonic tomography method, described in detail in [15–17], is used to test concrete floors that are up to 2500 mm thick on one side; to determine their thickness; to detect the distribution of reinforcement consisting, for example, of several layers of mesh; and to detect cracks invisible on the surface, foreign inclusions, air voids or areas filled with a liquid or material significantly different in physical properties from the surrounding concrete.	 multi-head antenna	Quantitative images of the concrete floor structure in which the numerical values of each pixel describe the speed of ultrasonic wave propagation in the concrete. An example of such an image is given below: 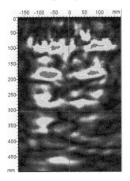

TABLE 5.4

Synthetic Description of Selected Non-Destructive Methods Useful for the Location of Reinforcement

Name and Description of the Method	View of the Apparatus Used	Recorded Parameters, Exemplary Test Results
The magnetic method is based on the analysis of the change in the magnetic field in the vicinity of rebar. This method is used to locate reinforcing bars and to determine their diameter and the thickness of their concrete cover. During the tests, the test apparatus is moved over the surface of the floor. In order to determine the class and grade of the reinforcement steel, a control outcrop should be made for this purpose, because it cannot be done with the non-destructive method [18–20].	measuring head	ϕ – rebar diameter, c – rebar cover. The location of the reinforcing bars is marked with chalk on the surface of the tested floor: 

TABLE 5.4 (*Continued*)
Synthetic Description of Selected Non-Destructive Methods Useful for the Location of Reinforcement

Name and Description of the Method	View of the Apparatus Used	Recorded Parameters, Exemplary Test Results
The radar method involves emitting electromagnetic waves with frequencies ranging from short to ultra-short radio waves, and then recording the waves reflected from layers that are characterized by variable dielectric properties [21]. The operating range of the radar method, in terms of the depth calculated from the surface of the floor, depends on the structure of the concrete, the type of radar antenna and the magnitude of the excited pulse frequencies. In typical devices, this range is up to 750 mm. The result of research on the location of reinforcement, carried out using the radar method, are the so-called phalograms, i.e., a record of all the reflected signals recorded during profiling (the probe travels over the surface of the floor). The image of the reinforcement on the phalogram is a distortion of the course of contour lines in the form of a hyperbola with arms pointing towards the bottom of the phalogram. The top of the hyperbola shows the location of the rebar. The test devices do not make it possible to explicitly determine the diameter of the reinforcement. However, it is possible to determine the diameter of the bars crossing the floor. For this purpose, a scan of the structure is made in two perpendicular directions over the crossing bars, and then the thickness of the concrete cover is measured. The difference between the thickness of the coverings of the crossing bars is the diameter of the bar located closer to the surface of the tested floor.	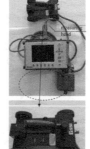	A sample image of the phalogram obtained during the tests described in [22]: 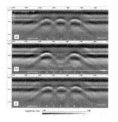

TABLE 5.5
Synthetic Description of Selected Methods Useful for the Assessment of Concrete Floor Surface Morphology

Name and Description of the Method	View of the Apparatus Used	Recorded Parameters, Exemplary Test Results
The laser triangulation method involves measuring the deformation of the line produced by the laser beam by a measuring device in the form of a special triangulation scanner. The measuring system of the scanner consists of a laser light source, a measuring object and a camera. In the scanner, a camera driven by a stepper motor is responsible for measuring the distance between individual points located on the tested surface. An original 3D laser scanner of a new design was developed to study the morphology of concrete surfaces, with dimensions of 50 × 50 mm, using the non-contact method and laser triangulation. This scanner is described in detail in [23]. A diode laser generator was used as the light source. In the case of laser triangulation testing of large areas, for example, industrial floors or floors in large-scale garages, where the research of the concrete surface morphology is carried out at many points, the testing process can be improved by using a driving platform. Such a platform is an element of a utility model [24]. It is made of a frame base to which, at its corners, electric motors and road wheels are attached. The platform is equipped with a remote control.	Laser scanner:  Driving platform: 	Sq – Root mean square height, Ssk – Skewness, Sku – Kurtosis, Sp – Maximum height of summits, Sv – Maximum depth of valleys, Sz – Maximum height, Sa – Arithmetic mean height, Smr – Areal material ratio, Smc – Inverse areal material ratio, Sxp – Extreme peak height, Sal – Auto-correlation length, Str – Texture-aspect ratio, Std – Texture direction, Sdq – Root mean square gradient, Sdr – Developed interfacial area ratio, Vm – Material volume, Vv – Void volume, Vmp – Peak material volume, Vmc – Core material volume, Vvc – Core void volume. The above parameters are listed in the ISO 25178 standard [25]. An exemplary isometric view of a surface tested with a laser scanner: 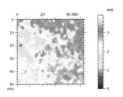

(continued)

TABLE 5.5 (*Continued*)
Synthetic Description of Selected Methods Useful for the Assessment of Concrete Floor Surface Morphology

Name and Description of the Method	View of the Apparatus Used	Recorded Parameters, Exemplary Test Results
The optical method is useful for assessing the morphology of a concrete floor surface and is based on the projection of fringes with white light. The measurement involves projecting these fringes, and then two cameras registering them simultaneously. Thanks to this, the measuring system determines the position of the sensor and then analyses the values of the measurement points in the coordinate system of the measured object. A typical measurement system in this method is equipped with two cameras that record light fringes forming a point cloud. The cameras assign three coordinates to each measurement point with an accuracy of 20 μm. Both cameras independently measure the refractions of the light fringes and perform the test again in the case of significant differences at the same test point.		Parameter values are recorded analogically to the laser triangulation method listed in the ISO 25178 standard [25].

TABLE 5.6
Synthetic Description of the Pull-off Method Useful in in Situ Research

Name and Description of the Method	View of the Apparatus Used	Recorded Parameters, Exemplary Test Results
In the **pull-off method**, the measurement of the near-surface tensile strength or interlayer pull-off off adhesion is made by measuring the maximum pull-off force value of the metal discs glued to the concrete floor surface with an actuator that has a digital or dial gauge. For this purpose, holes are made in the concrete floor of the surface layer with a diameter of 50 mm and a depth of at least 5 mm, and steel discs are glued and torn off the surface.		F_b – maximum strength of the steel disc breaking off the concrete floor, f_b – interlayer pull-off adhesion, f_h – subsurface tensile strength.

TABLE 5.7
Synthetic Description of Selected Non-Destructive Methods Useful in Laboratory Tests

Name and Description of the Method	View of the Apparatus Used	Recorded Parameters, Exemplary Test Results
The X-ray tomography method allows the reconstruction of a three-dimensional image of a concrete sample taken from the floor under study. The image is based on two-dimensional projections obtained during the floor's scanning with an X-ray beam. This method makes it possible to determine many of the parameters that characterize the concrete structure, including porosity, the distribution of pores in terms of size, the distribution of aggregate grains, the distribution of microdamages, the distribution of dispersed reinforcement, the degree of carbonation of concrete, and also the determination of the value of the modulus of elasticity. The test results obtained with this method can be very useful, for example, in the interpretation of the causes of structural damage.	 	μ – material x-ray absorption coefficient. An exemplary view of the structure of a tested concrete sample: 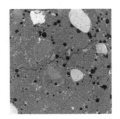
Optical microscopic methods involve illuminating the surface of the tested concrete sample, which is taken from the floor, at an angle α, usually equal to 45°, through a narrow slit. This method uses the principles of geometric optics, and on this basis, the height and spacing of unevenness on the tested surface are determined. The advantage of this method is the lack of direct contact between the measuring equipment and the tested concrete surface. The disadvantages include the limitation of the sample size by the size of the chamber of the device and the dependence of the measurement on the scattering of light by the material of the measured surface.		According to [29], values of the following parameter can be obtained: A_A – share of exposed aggregate visible on the surface. Optical view of an analysed concrete sample: 

5.8 CONCLUSIONS

This chapter contains synthetic information on selected non-destructive and semi-destructive methods. The choice of methods was guided primarily by their application in the case studies presented in Chapters 8–14. The information provided includes the name and a very brief description of the method, a view of the apparatus used in a given method, specification of the recorded parameters supported by exemplary results from own research.

REFERENCES

1. Zybura, A., Jaśniok, M., & Jaśniok, T. (2017). Diagnostics of reinforced concrete structures (in Polish). Warsaw: Polish Scientific Publishers PWN.
2. Bungey, J.H., Millards, S.G., & Grantham, M.G. (2006). Testing of concrete in structures. Boca Raton, FL: CRC Press.
3. Malhotra, V.M., & Carino, N.J. (2003). Handbook on nondestructive testing of concrete. Boca Raton, FL: CRC Press.
4. Drobiec, Ł., Jasiński, R., & Piekarczyk, A. (2010). Diagnostics of reinforced concrete structures (in Polish). Warsaw: Polish Scientific Publishers PWN.
5. Brunarski, L., & Dohojda, M. (2015). Diagnostics of concrete strength in a structure (in Polish). Warsaw: Building Research Institute.
6. Breysse, D., & Martínez-Fernández, J.L. (2014). Assessing concrete strength with rebound hammer: Review of key issues and ideas for more reliable conclusions. Materials and structures, 47(9), 1589–1604.
7. Runkiewicz, L., & Sieczkowski, J. (2018. Assessment of concrete strength in a structure based on sclerometric tests. Guide (in Polish). Warsaw: Building Research Institute.
8. EN 12504-2:2013-03. Testing concrete in structures - Part 2: Non-destructive testing - Determination of rebound number.
9. Nieświec, M., & Sadowski, Ł. (2021). The assessment of strength of cementitious materials impregnated using hydrophobic agents based on near-surface hardness measurements. Materials, 14(16), 4583.
10. Brunarski, L., & Runkiewicz, L. (1997). ITB instruction no. 209. instructions for the use of the ultrasonic method for non-destructive quality control of concrete in a structure (in Polish). Warsaw: Building Research Institute.
11. EN 12504-4:2005. Testing concrete - Part 4: Determination of ultrasonic pulse velocity.
12. EN 13791:2008. Assessment of in-situ compressive strength in structures and precast concrete components
13. Hoła, J., Bień, J., Sadowski, Ł., & Schabowicz, K. (2015). Non-destructive and semi-destructive diagnostics of concrete structures in assessment of their durability. Bulletin of the Polish Academy of Science. Technical sciences, 63(1), 87–96.
14. Sansalone, M., & Streett, W.: Impact – echo. (1997) Nondestructive evaluation of concrete and masonary. Ithaca: Bullbrier Press.
15. Bishko, A.V., Samokrutov, A.A., & Shevaldykin, V.G. (2008). Ultrasonic echo-pulse tomography of concrete using shear waves low frequency phased antenna arrays. e-Journal of non-destructive testing & ultrasonics, 13, 1–9.
16. Kozlov, V.N., Samokrutov, A.A., & Shevaldykin, V.G.: Thickness measurements and flaw detection in concrete using ultrasonic echo method. Journal of Nondestructive testing and evaluation, 1997, 13, 73–84.
17. Hoła, J., & Schabowicz, K. (2010). State-of-the-art. Nondestructive methods for diagnostic testing Of building structures – Anticipated development trends. Archines of civil and mechanical engineering, 10(3), 5–18.
18. Drobiec, Ł., Determining the parameters of reinforcing steel in a structure (in Polish). XXIX National Workshop of Structural Designers. Szczyrk, March 26–29, 2014 pp. 181–256.

19. Drobiec, Ł., Jasiński, R., & Piekarczyk, A., Localization of defects in structures and reinforcing steel – methods (in Polish), 21st Polish National Conference Workshop for Structural Designers, Szczyrk, March 8–11, 2006, pp. 133–208.
20. Drobiec, Ł., Jasiński, R., & Piekarczyk, A. (2007) Methods of locating reinforcing steel in reinforced concrete structures. The electromagnetic method (in Polish)). Building review, 12, 31–37.
21. Conyers, L., & Goodman, D. (1997). Ground – Penetrating radar. Walnut, Creek, CA: AltraMira Press.
22. Pospisil, K., Manychova, M., Stryk, J., Korenska, M., Matula, R., & Svoboda, V. (2021). Diagnostics of reinforcement conditions in concrete structures by GPR, impact-echo method and metal magnetic memory method. Remote sensing, 13(5), 952.
23. Czarnecki, S., Hoła, J., & Sadowski, Ł., A nondestructive method of investigating the morphology of concrete sur-faces by means of newly designed 3D scanner. XI European Nondestructive Testing Conference "ECNDT 2014", Prague, Czech Republic
24. Żelazny, Z., Sadowski, Ł., Kupczyk, M., Czarnecki, S., Hoła, J., & Wrzecioniarz, P., Utility model. Poland, no. 70494. A mobile platform for a device for examining the morphology of flat surfaces, especially concrete ones (in Polish): Int. Cl. B62D 39/00, B62D 25/02. Application no 126268 z 14.04.2017. Published 31.01.2019.
25. EN ISO 25178: Geometric Product Specifications (GPS) - Surface texture: areal
26. EN 12504-3:2006. Testing concrete in structures - Part 3: Determination of pull-out force.
27. Malhotra, V.M., & Carino, N.J. (2004). Handbook on nondestructive testing of concrete. Boca Raton, FL: CRC Press.
28. ATM C 900-82: Standard Method for Pullout Strength of Hardened Concrete, ASTM Committee on Standards, 1916 Race Street, Philadelphia, Pa. 19103, USA.
29. Sadowski, Ł., Żak, A., & Hoła, J. (2018). Multi-sensor evaluation of the concrete within the interlayer bond with regard to pull-off adhesion. Archives of civil and mechanical engineering, 18(2), 573–582.

6 The Reliability of Non-Destructive Methods in Floor Diagnostics

6.1 LIMITATIONS AND CONDITIONS OF USE FOR SELECTED METHODS

6.1.1 SCLEROMETRIC METHOD

Despite the fact that the sclerometric method is well known and widely used, it was decided to highlight the most important limitations and conditions of its use. In the case of the sclerometric method, in which the basic research device is the N-type Schmidt hammer, it should be taken into account that:

- With the Schmidt N-type hammer, the compressive strength f_c of the concrete is not directly determined – but instead the rebound R number. This number is a measure of the dynamic hardness of the concrete – therefore the empirical relationship between f_c and R in the form $f_c = f_c(R)$. According to [1–3], it should be noted that there is no universal such relationship for all concretes.
- The $f_c - R$ relationship is influenced by many different factors, such as the age of the concrete, the moisture in the concrete, stresses in the concrete, geometry and shape of the test pieces under test, the position of the hammer in relation to the level during the test, etc. In order to correctly assess the strength, these factors should be taken into account by introducing appropriate correction or correction factors, either to the rebound number or to the strength [4, 5]. The influence of some of the above-mentioned factors is very important for a reliable assessment, which is demonstrated below:
  - Age influence – this is mainly related to the carbonation phenomenon progressing over time, and the resulting calcium carbonate "hardening" the concrete structure, in turn causing an uneven distribution of concrete strength. As shown in Figure 6.1, based on [6], as the distance from the surface of the concrete element increases, the rebound number R decreases its value more than the value of the compressive strength f_c. The influence of age is also manifested in the number of crystalline deposits in the concrete structure, which increase with age.

As can be seen in Figure 6.2, based on [6], the greatest cumulative effect of carbonation and crystallization on the hardness of concrete in the near-surface layer of a floor

DOI: 10.1201/9781003288374-6

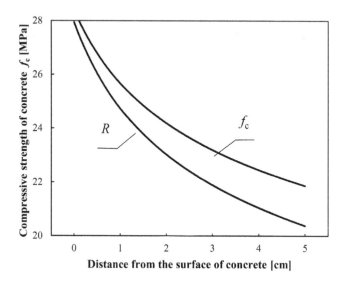

FIGURE 6.1   An exemplary graph of the distribution of the rebound number R and the concrete strength f_c with regards to the distance from the concrete surface.

occurs in the first year. The ratio of the concrete strength in the deeper layers f_{cint} to the subsurface strength f_{cext} stabilizes over time at about 0.6.

- The influence of humidity – this manifests itself in the deterioration of the dynamic hardness of concrete and a decrease in the rebound number. The scale of this phenomenon is shown in Figure 6.3, based on the data contained in [6–8].

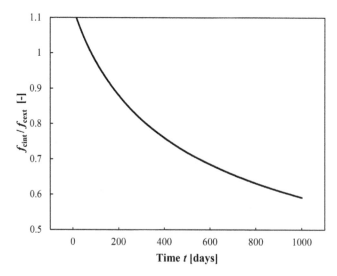

FIGURE 6.2   An exemplary diagram of the dependence of the change in the quotient of concrete strength in the deeper layers f_{cint} of the floor and in the subsurface layers f_{cext} of the floor on the time t.

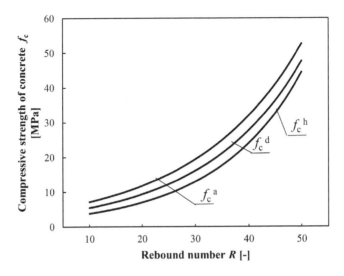

FIGURE 6.3 Empirical relationships $f_c - R$ for the Schmidt N-type hammer in tests of ordinary concrete (f_c^a – air-dry concrete, f_c^d – dry concrete, f_c^h – wet concrete).

- The influence of the geometry and shape of the tested floor, including the scale effect – this is clearly noticeable as the strength of the concrete increases, hence the empirical relations for the same concrete. However, for different types of test elements, they will differ, as shown in Figure 6.4 – developed on the basis of the data contained in [6–8].
- The positioning of the Schmidt hammer during the diagnostics of concrete structures – in a direction other than horizontal is acceptable, but

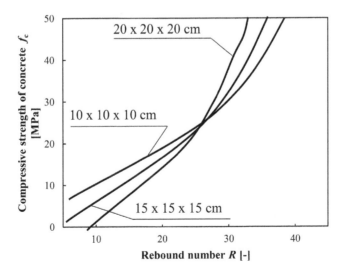

FIGURE 6.4 Empirical relationships $f_c - R$ for the Schmidt N-type hammer for different dimensions of the sample elements.

TABLE 6.1

Comparison of the Requirements for Measurement Sites on the Floor Surface According to the Standards EN-12504-2:2002 [5] and PN-B-06262 [9]

Type of Requirement	EN-12504-2:2002 [5]	PN-B-06262 [9]
The area of the measuring site	About 900 cm²	About 50 cm²
Minimum number of measurement sites	No requirements	12
The minimum number of measurement points at a given measurement site	9	5
Minimum distance between measurement points	25 mm	20 mm

then the correction factor ΔR for the rebound number R must be introduced. In the case of concrete floors, this is illustrated in Figure 6.5, based on the data contained in [4].

- The minimum size of the measurement site (area) and the distribution of the measurement points on the floor surface, according to [5, 9], should be as specified in Table 6.1.
- The thickness of the diagnosed concrete floor is limited. In the case of the N-type Schmidt hammer, according to [5–10], it should not be smaller than 100 mm. Moreover, according to [10], it should not be greater than 200 mm, but may be up to 400 mm, with two-sided access interpreted as from the top and side of the floor.
- Selection of test sites – this is important for the reliability of the sclerometric measurements. Test sites should be evenly distributed over

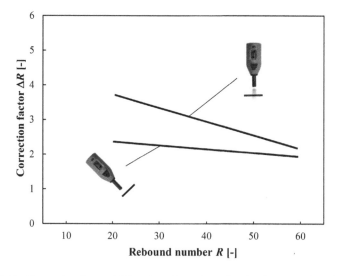

FIGURE 6.5 Correction factor ΔR for the rebound number in the case of positioning the N-type Schmidt hammer in a direction other than horizontal.

the entire surface of the floor in order to be representative for it. They should not be located less than 30 mm from the edge, should not be on the line of reinforcement located in the concrete at a depth of less than 30 mm, and should not be located in places where thick aggregate grains are visible right next to the surface.

- In the early stage of concrete maturation, or where the compressive strength of concrete in the floor is less than 7 MPa, the use of the sclerometric method is limited – the rebound numbers are too low, and do not ensure the accuracy of the reading. In addition, it causes undesirable damage to the floor surface during tests, which then requires repair [11]. As stated in [5], the minimum age of the tested concrete should not be less than 3 days, but in the case of high-performance concretes, this restriction does not apply.

6.1.2 Ultrasonic Method

In the case of the ultrasonic method, it is worth considering the following considerations with regards to floor diagnostics:

- When using the ultrasonic method, the compressive strength f_c of the concrete is not directly determined, but instead the velocity of the longitudinal ultrasonic wave c_L. In practical applications, in analogy to the sclerometric method, an empirical relationship between f_c and c_L should each time be established in the form $f_c = f_c (c_L)$. This relationship is established with the exact or simplified method.
- There is no universal relationship between the velocity of the longitudinal ultrasonic wave c_L and the compressive strength of the concrete f_c, which is representative for all concretes.
- The $f_c - c_L$ relationship is influenced by many different factors, resulting not only from the composition and technology of execution and structure of the concrete floor, but also from the operating conditions of the diagnosed floor. These are, among others: the type of aggregate, coarse aggregate grain size, grain size distribution, moisture of the concrete, conditions for concrete maturation, the presence of reinforcement at the test site, and the level of stresses in the concrete. They are important when determining the empirical dependence, which is most often done with the use of test elements taken from the floor for which this dependence is established. For some factors, for example, concrete moisture or the influence of reinforcement on the results of sclerometric tests (which are already identified and described in literature), correction factors are used. These correction factors are approximated below:
  - The increase in concrete moisture is manifested by an increase in the ultrasonic wave velocity c_L. The scale of this phenomenon is explained in Figure 6.6, based on the data contained in [12].

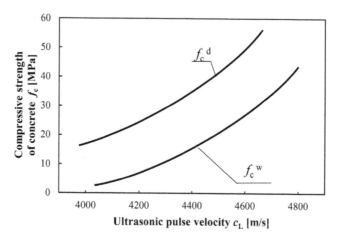

FIGURE 6.6 Influence of concrete moisture on the longitudinal ultrasonic wave velocity c_L (f_c^d – dry concrete, f_c^w – wet concrete).

- The influence of the reinforcement on the ultrasonic wave velocity depends on the amount and location of the reinforcement in the measuring section. It is recommended to avoid areas of the floor where the measurement is taken parallel to the rebar. In order to determine the location of the reinforcement, supporting tests, performed with the use of non-destructive methods suitable for this purpose, are necessary. For example, Figure 6.7, based on the data presented in [7], shows the values of the correction factor while taking into account the effect of the length of reinforcing bars l_s perpendicular to the direction of the propagation of the ultrasonic wave in the case of testing an element with a length of l.

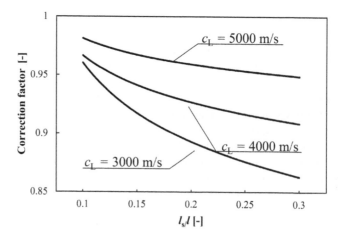

FIGURE 6.7 Values of the correction factor taking into account the effect of the length of reinforcing bars l_s perpendicular to the direction of the propagation of the ultrasonic wave in the case of testing an element with a length l.

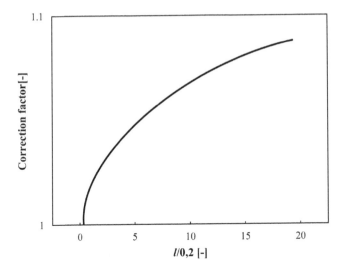

FIGURE 6.8   The value of the correction coefficient reducing the ultrasonic wave velocity measured on different bases.

On the other hand, [13], shows how the ultrasonic wave velocity in a tested element (c_{Le}) depends on the diameter of the reinforcement (d) and the velocity of the ultrasonic wave in the concrete without taking into account the reinforcement (c_{Lc}):

$$c_{Le} = 5.9 - 10.4\frac{5.9 - c_{Lc}}{d} \qquad (6.1)$$

Relation (6.1) is valid for ultrasonic heads with a frequency of 54 kHz.

- The influence of the measurement path length is presented in detail in [14]. Its minimum length should be 100 mm if the maximum grain size of the coarse aggregate in the concrete is 20 mm. As the length of the measuring path increases, the ultrasonic wave velocity decreases c_L. If the length of the path exceeds 600 mm, then on the basis of [7], c_L must be corrected by the value of the correction factor that can be read from Figure 6.8.

Moreover, according to [12], the measurement path length should be greater than the ultrasonic wavelength. The size of the coarse aggregate must also be smaller than the ultrasonic wavelength in order to avoid a reduction of the wave energy and possible signal loss at the receiver. In Table 6.2, on the basis of the data contained in [12], restrictions on the measurement path length in the ultrasonic method are summarized.

6.1.3   ULTRASONIC TOMOGRAPHY METHOD

Currently, according to [15], in the case of the ultrasonic tomography method, it is worth considering that:

TABLE 6.2

Basic Restrictions Regarding the Measurement Path Length in the Ultrasonic Method

Recommended Measurement Path Length (for Coarse Aggregate with a Maximum Diameter of 20 mm)	The Smallest Side Dimension or the Maximum Size of Coarse Aggregate (mm)	
	For ultrasonic wave velocity of 3800 m/s	For ultrasonic wave velocity of 4600 m/s
100 mm – 10 m for ultrasonic heads with a frequency of 54 kHz	70	85
100 mm – 3 m for ultrasonic heads with a frequency of 82 kHz	46	56

- The minimum thickness of a tested concrete floor is limited to 50 mm, while the maximum thickness is practically not limited, and may be up to 2500 mm.
- The researcher's experience is important for the credibility of the obtained results. The relatively small number of research cases to date means that in some cases the correct interpretation of the obtained images may still be difficult for some researchers. With the passage of time, the importance of this will decrease, which is due to the fact that ultrasonic tomographs are more and more widely available and the number of their applications is growing, resulting in more experience being gained.

6.1.4 IMPACT-ECHO AND IMPULSE RESPONSE METHODS

The most important remarks concerning the impact-echo and impulse response methods, from the point of view of the reliability of the obtained results, are as follows:

- Both methods are not applicable when the structure is influenced by mechanical noise or high amplitude electrical noise.
- It is not possible to assess the depth of cracks if they are partially or completely filled with water.
- According to [16], a reliable assessment of the thickness of a layered concrete floor is only possible when there is a sufficient difference in acoustic impedance or if there are enough air voids between the overlay and the substrate to produce a measurable reflection from the bottom of the floor.
- In the impact-echo method, the inductor must be positioned at a distance from the transducer that is less than 0.4 times the thickness of the tested floor.

- The thickness of the tested floor is not limited, because according to [17] the thickness of the element tested with this method can be up to 1 m.
- Due to the edge effect, the tests should not be carried out closer than 300 mm from the edge of the floor.

6.1.5 METHODS OF ASSESSING HUMIDITY

Non-destructive humidity testing methods have certain limitations and conditions that make their use in some situations problematic, with the obtained results not being very reliable:

- Not all methods can quantitatively evaluate the moisture content and its distribution in a concrete floor, and instead they assess these features only qualitatively. This applies to the chemical method of indicator papers, in which the moisture is determined graphically on the basis of the colour change of the papers as a result of contact with the surface of the moist material. The same applies to methods based on the measurement of thermal, optical and videography properties. They are only useful for locating wet areas, and do not have the possibility of determining the humidity values of individual places. This is a significant limitation.
- In the case of the most commonly used electrical, dielectric, resistive, and microwave methods in construction practice, an important condition is the need to scale the test equipment on a given object (accurate or simplified) in order to determine the correlation relationship between concrete moisture and the dimensionless parameter measured in a given method.
- The assessment of humidity by the dielectric or resistance method can be reliable only for the subsurface zone of the tested concrete floor with a thickness of about 60 mm, because this is the test range of these methods. For the microwave method, this condition practically does not exist, because its test range is up to about 300 mm.

6.2 THE COMPLEMENTARITY OF NON-DESTRUCTIVE TESTING

Limitations and conditions of using individual non-destructive methods for the purpose of diagnosing concrete floors mean that, in order to increase the credibility of tests conducted with these methods the most rational direction is to use at least two non-destructive methods simultaneously [18, 19]. If there is no such possibility, but it is possible to determine several parameters with a given method, then this option should be used. According to [20], several possible variants of such an approach can be distinguished, and are described briefly below (Figure 6.9).

It is important to note here that the current trends in the development of non-destructive methods of the diagnostics of concrete floors are focused on the automation of tests carried out with specific methods. Work is also underway on the

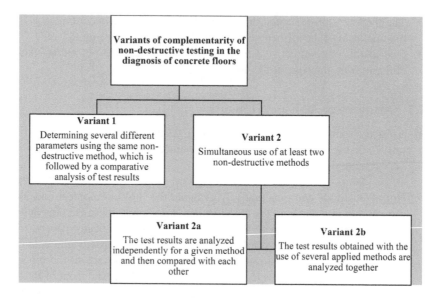

FIGURE 6.9 Variants of the approach to the complementarity of non-destructive testing in the diagnostics of concrete floors.

construction of special scanners that facilitate and accelerate the performance of tests using several non-destructive methods [21–23].

6.2.1 VARIANT 1

An example of the use of several different parameters determined by the same non-destructive method can be the ultrasonic method. These parameters are the velocity and amplitude of the longitudinal ultrasonic wave.

Another example of this would be the hammer method, where the following parameters can be used:

- the frequency corresponding to the ultrasonic wave reflected from the defect f_D,
- the frequency corresponding to the ultrasonic wave reflected from the bottom f_T,
- the velocity of an elastic longitudinal ultrasonic wave C_p.

Another example is the impulse response method, which makes it possible to obtain the values of the following five parameters:

- average mobility N_{av},
- dynamic stiffness K_d,
- mobility slope M_p/N,
- average mobility times mobility slope $N_{av} M_p/N$,
- voids index v.

6.2.2 VARIANT 2

An example of the use of at least two non-destructive methods are the ultrasonic and sclerometric methods used to determine the compressive strength of concrete. The test results obtained with these methods are most often assessed independently and then compared with each other. In the past, attempts were made to apply empirical relationships for two variables and to statistically analyse the results obtained independently with individual methods [24]. Such a combination concerned the ultrasonic and sclerometric methods. Although it allowed the average relative error of the strength assessment to be significantly reduced, it required, a new empirical relationship to be each time developed for the tested concrete [25].

An example of the use of two non-destructive methods are the impulse response method and the impact-echo method for assessing delamination in layered concrete floors [26]. In this case, the impulse response method was suitable for a rough search for areas of the floor in order to identify fragments in which delamination is likely to occur. In turn, the impact-echo method was used to accurately locate the depth of delamination in the floor.

6.3 THE IMPORTANCE OF CONTROL SECTIONS

The results of non-destructive diagnostics of concrete floors should be, if possible, verified each time by making control sections that give credibility to non-destructive testing.

They should also give the expert who diagnoses the concrete floor a more complete basis for a more precise assessment of the destructive processes and the assessment of safety in use, or more precise conclusions regarding the floor's expected durability. In some situations, for example those indicated below, making a control section (or sections) is highly advisable, or even necessary, for not only confirming the non-destructive results, but also for supplementing the information obtained in a non-destructive manner as highlighted in Examples 6.1, 6.2, and 6.3 provided below.

Example 6.1

An example of one such situation is non-destructive testing aimed at determining the location of rebars in the floor and the thickness of their concrete cover. This can be done non-destructively with different levels of accuracy. This can be confirmed by a control section (Figure 6.10), and then the following can be further assessed:

- the accuracy of the measurements made in relation to that provided by a given method, or the measurement with a given measuring device, could be disturbed by such factors as, for example: the non-parallel location of reinforcing bars in relation to the surface of the concrete floor, the presence of dispersed steel reinforcement in the concrete or aggregate with magnetic properties, or reinforcement corrosion,
- the type of reinforcing steel in terms of its ribbing and cross-section of reinforcing bars,
- the size of corrosion losses in the steel reinforcement.

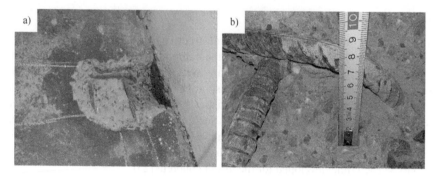

FIGURE 6.10 An exemplary view: (a) control section of reinforcement, which allows the diameter of the bars and cover thickness to be measured and type of steel, as well as the size of corrosion losses to be assessed, (b) close-up of the control section.

Example 6.2

Another example is non-destructive testing to determine the crack depth of a concrete floor. This can be done using acoustic methods (impact-echo or ultrasonic), but it is not certain whether the accuracy of the measurement is not disturbed by the presence of water in the crack. Since the width of the cracks is usually not the same along their entire height (thickness of the floor), and that they are hairline closer to the tip, they can often be filled with water in this zone (but not only there). Making a control section by taking a core sample along the crack line clearly explains this doubt and clarifies the non-destructive measurement (Figure 6.11).

Example 6.3

One more example is non-destructive testing aimed at determining the mass moisture content of concrete in the floor. It should be remembered that the range of

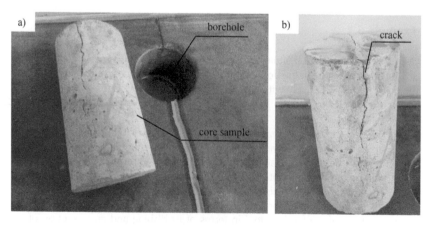

FIGURE 6.11 An exemplary view of: (a) a 50 mm diameter core sample taken along the crack line, (b) close-ups of the core sample.

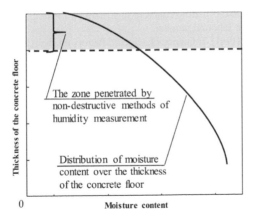

FIGURE 6.12 Typical distribution of moisture content along the thickness of a concrete floor.

non-destructive methods is much shorter and amounts to about 60 mm in the case of dielectric and resistive methods. If this information is compared with Figure 6.12, prepared on the basis of [27], which shows a typical distribution of mass moisture over the thickness of the concrete element, it becomes clear that in the case of a thick floor it is advisable to perform a control drilling through the floor in order to collect concrete samples for testing concrete moisture with the gravimetric method in the laboratory [28].

6.3 CONCLUSIONS

This chapter is devoted to highlighting the problem of the reliability of the use of non-destructive methods in the diagnosis of concrete floors. It has been clearly stated that the *sine qua non* conditions for obtaining reliable results with these methods are, firstly, the knowledge and consideration of the limitations and conditions for the use of non-destructive methods in the research, and secondly, their complementary use and, thirdly, verification of the test results obtained with non-destructive control sections. For a few selected methods, the most important constraints and conditions are given, and possible compatibility variants are also given.

REFERENCES

1. Malhotra, V.M., & Carino, N.J. (2003). Handbook on nondestructive testing of concrete. Boca Raton, FL: CRC Press.
2. Hoła, J., & Runkiewicz, L. (2018). Methods and diagnostic techniques used to analyse the technical state of reinforced concrete structures. Structure and environment, 10, 309–337.
3. Skramtaev, B.G., & Leshchinszy, M.Y. (1966). Complex methods of non-destructive tests of concrete in construction and structural works, RILEM Bull. (Paris), New Series No. 30, 99.
4. ITB 210/1977 manual. Instructions for the use of Schmidt hammers for non-destructive quality control of concrete (in Polish)

5. EN 12504-2:2013-03. Testing concrete in structures - Part 2: Non-destructive testing - Determination of rebound number.
6. Runkiewicz, L. (1991). Influence of selected factors on the results of sclerometric tests of concrete (in Polish). Warsaw: Scientific works of the Building Research Institute.
7. Drobiec, Ł., Jasiński, R., & Piekarczyk, A. (2010). Diagnostics of reinforced concrete structures (in Polish). Warsaw: Polish Scientific Publishers PWN.
8. Flaga, K. (1970) Influence of the size of test elements on the evaluation of concrete strength using the sclerometric method (in Polish). Inżynieria i Budownictwo, no 10, pp. 395–399
9. PN-74/B-06262 Non-destructive testing of concrete structures - Sclerometric method - Tests of concrete compressive strength with a Schmidt N-type hammer (in Polish)
10. Runkewicz, L., & Sieczkowski, J. (2018). Assessment of concrete strength in a structure based on sclerometric tests. Guide (in Polish). Warsaw: Building Research Institute.
11. Mitchell, L.J., & Hoagland, G.G. (1961) Investigation of the impact tube concrete test hammer, Bull. No. 305, Highway Research Board, 14.
12. Bungey, J.H. (1980). The validity of ultrasonic pulse velocity testing of in-place concrete for strength. NDT International, 13(6), 296–300.
13. Chung, H.W. (1978). 'Effects of embedded steel bars upon ultrasonic testing of concrete'. Magazine of Concrete Research, 30, No. 102, March, 19–25.
14. Stawiski, B. (1988) Badania niektórych materiałów i konstrukcji budowlanych wybranymi metodami nieniszczącymi. Centrum Usług Techniczno-Organizacyjnych Budownictwa. Ośrodek we Wrocławiu, Wrocław.
15. Hoła, J., Bień, J., Sadowski, Ł., & Schabowicz, K. (2015). Non-destructive and semi-destructive diagnostics of concrete structures in assessment of their durability. Bulletin of the Polish Academy of Science. Technical sciences, 63(1), 87–96.
16. ASTM, C. (2000). Test method for measuring the P-wave speed and the thickness of concrete plates using the impact-echo method. Annual Book of ASTM Standards, 1383(04.02).
17. ASTM C1740-10, "Evaluating the condition of concrete plates using the impulse response method." Standard practice. West Conshohocken, PA: ASTM International.
18. Breysse, D. (2012). Nondestructive evaluation of concrete strength: An historical review and a new perspective by combining NDT methods. Construction and building materials, 33, 139–163.
19. Ali-Benyahia, K. Sbartaïc, Z.M., Breyssec, D., Kenaid, S., Ghricia, M. (2017) "Analysis of the single and combined non-destructive test approaches for on-site concrete strength assessment: General statements based on a real case-study." Case studies in construction materials 6, 109–119.
20. Adamczewski, G., Garbacz, A., Piotrowski, T., & Załęgowski, K. (2013). Application of complementary NDT methods in the diagnosis of concrete structures (in Polish). Building materials.
21. Reichling, K., Raupach, M., Wiggenhauser, H., Stoppel, M., Dobmann, G., & Kurz, J.: BETOSCAN-Robot controlled non-destructive diagnosis of reinforced concrete decks, 7th International Symposium on Non Destructive Testing in Civil Engineering 2009, Nantes, France.
22. Leibbrandt, A., Caprari, G., Angst, U., Siegwart, R.Y., Flatt, R.J., & Elsener, B. (2012, September). Climbing robot for corrosion monitoring of reinforced concrete structures. In 2012 2nd International Conference on Applied Robotics for the Power Industry (CARPI) (pp. 10–15). IEEE.
23. Garbacz, A., Piotrowski, T., Zalegowski, K., & Adamczewski, G. (2013). UIR-scanner potential to defect detection in concrete. Advanced materials research, 687, 359–365.
24. Nagrodzka-Godycka, K. (1999). Testing the properties of concrete and reinforced concrete in laboratory conditions (in Polish). Warsaw: Arkady.
25. Runkiewicz, L. (2002). Research of reinforced concrete structures (in Polish). Warsaw: Gamma office.

26. Hola, J., Sadowski, L., & Schabowicz, K. (2011). Nondestructive identification of delaminations in concrete floor toppings with acoustic methods. Automation in construction, 20(7), 799–807.
27. Hoła, J. (2018). Degradation of historic buildings as a result of excessive moisture - selected problems (in Polish). Construction and architecture, 17(1), 133–148.
28. Hussain, A., & Akhtar, S. (2017). Review of non-destructive tests for evaluation of historic masonry and concrete structures. Arabian journal for science and engineering, 42(3), 925–940.

7 The Calibration and Scaling of Test Equipment

7.1 THE CALIBRATION OF TEST EQUIPMENT

To ensure the required measurement accuracy, the test equipment used should be calibrated in accordance with the recommendations of the manufacturer of the equipment, usually before each use or at regular intervals [1].

The terms "calibration" and "scaling" are often used interchangeably in non-destructive testing. In ultrasonic flaw detectors, they are usually scaled using reference or reference samples [2–6]. Therefore, in ultrasonic testing, there is a "time base scaling" performed using a reference sample (Figure 7.1).

In turn, in the case of the sclerometric method, calibration is performed using hardness standards with calibration certificates. Typically, a calibration anvil is used for this purpose, on which, with an N-type sclerometer, an average rebound number R of 80±2 should be obtained [7].

Calibration is a set of operations that establish, under certain conditions, relations between the values of the measured quantity indicated by the test device, or the values represented by the standard of measurement or reference material, and the appropriate values of the quantity realized by the standards of the measurement unit.

On the other hand, scaling (of test equipment) is the determination of the position of indications of a measuring instrument, depending on the appropriate value of the measured quantity.

In the case of defectoscopic tests, the accuracy and reliability of test results are significantly influenced by the test equipment and apparatus, probes and test heads, basic test materials, standard samples and reference samples. Of this group, only reference and reference samples can be calibrated. On the other hand, the test equipment and apparatus, probes and test heads, and basic test materials cannot be "calibrated" due to the fact that the connection with the standards of measurement units in their case does not apply.

Conversely, test equipment used directly in defectoscopic tests, i.e., measures, calipers, lux meters, densitometers, etc., as well as instruments for measuring environmental conditions – should be calibrated if the uncertainty of the test result related to their calibration is a fraction of the total uncertainty of the test.

7.2 PRECISE SCALING

The presented procedure for expressing uncertainty with fine scaling is adapted to the specificity of correlation. The specificity of the relationship is a small sample population from a statistical point of view. However, the main task here is not to assess the existence of correlation relationships, but only to determine the regression function (equation) [8–11].

DOI: 10.1201/9781003288374-7

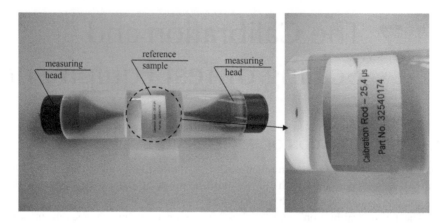

FIGURE 7.1 Reference sample used to scale the test equipment for ultrasonic testing.

The input data for the regression analysis using the fine scaling method is a set of pairs of X and Y results obtained at individual measuring sites (a minimum of 18 pairs is recommended in accordance with [1] for compressive strength, and a minimum of 30 for moisture). The indicated regression functions are most often expressed in the form of a simple equation:

$$f(Y) = a_1 + a_2 Y,$$ (7.1)

or parabola equation:

$$f(Y) = a_1 + a_2 Y + a_3 Y^2.$$ (7.2)

The parameters a_1, a_2, and a_3 (appropriately nominated numbers) of the regression function are determined by the method of the least squares of deviations of the marked Y values from the theoretical values (calculated from regression equations) corresponding to the $\hat{Y}$. These parameters can be determined in the traditional way or with the use of available numerical programs. For example, in accordance with [1], in the case of the sclerometric method, these pairs of results represent the values of the compressive strength f_c of the concrete on the core samples with a diameter of 100 mm, and the determination of the rebound numbers R – reduced to readings with a horizontal sclerometer setting (angle of inclination $\alpha = 0$).

An important issue is the fit (reliability) of the determined regression $f(Y)$, and in particular the fulfilment of the following requirements:

- calculation of the standard error of estimation, determining the confidence limits of the fit, and the limits of the single result tolerance,
- the correlation relationship should be determined assuming the possibility of a 10% underestimation of the strength.

TABLE 7.1

Exact Scaling for Determining the Correlation Relationship $f(Y) - Y$ in Tests Aimed at Determining the Compressive Strength of Concrete and the Bulk Moisture of Concrete in Floors

The Type of Feature to be Determined	Selection of the Relationship $f(Y) - Y$ by Means of Correlation Analysis
Compressive strength of concrete f_c	It is required to collect a minimum of 18 core samples of concrete from the structure for testing in order to obtain a minimum of 18 pairs of $f_c - R$ results, where R is the rebound number determined using a sclerometer
Bulk moisture of concrete U_m	It is required to collect a minimum of 30 concrete samples from the structure for testing in order to obtain a minimum of 30 pairs of $U_m - X$, where X is a unitless number determined with a moisture meter

In other words, the appropriate correlation relationship is a curve or a straight line that limits the confidence area of the plot from the bottom of the designated function $f(Y)$ with a significance level of 10%. According to [1], such a dependence then provides a level of safety, at which it is expected that 90% of the strength value will be higher than the value determined from this relationship.

The standard deviation of the residuals (the correct name for the standard error of the estimate), i.e., the deviations of the marked values Y, i.e., the deviations of the determined values from the theoretical values (calculated from regression equations) corresponding to the value of $\hat{Y}$, in the case of one explanatory variable, is expressed by the formula:

$$S = \sqrt{\frac{1}{n-2}\sum_{i=1}^{n}\left(Y - \hat{Y}\right)^2}. \tag{7.3}$$

For example, in the case of sclerometric testing, fine scaling determines the correlation between the rebound number R, and the compressive strength f_c. Table 7.1 shows the methods of determining the correlation relationship $f(Y) - Y$ in tests aimed at determining the compressive strength of concrete and the bulk moisture of concrete.

7.3 APPROXIMATE SCALING

Approximate scaling is based on the adoption of hypothetical ready base (basic) regression curves that are prepared, for example, from the literature. They are then shifted based on the test results of the marked values at selected measurement points. Regardless of the method of determining this dependence, it is necessary to take a

TABLE 7.2

Approximate Scaling for Determining the Correlation Relationship $f(Y) - Y$ in Tests to Determine the Compressive Strength of Concrete and the Bulk Moisture of Concrete in Floors

The Type of Feature to be Determined	Selection of the Hypothetical Relationship $f(Y) - Y$ from Among Those Available in the Literature
Compressive strength of concrete f_c	It is required to collect a minimum of 9 concrete samples from the structure for testing in order to obtain a minimum of 9 pairs of results $f_c - R$, where R is the rebound number determined using a sclerometer
Concrete bulk moisture U_m	It is required to collect a minimum of 6 concrete samples from the structure for testing in order to obtain a minimum of 6 pairs of results $U_m - X$, where X is a unitless number determined with a moisture meter

certain minimum number of samples from the tested element for destructive testing, which is specified in Table 7.2.

For scaling the underlying type regression curve $f(Y) = a_1 + a_2 Y + a_3 Y^2$ presented in Equation (7.2), a sample population is required with a specific, depending on the algorithm used, number of n pairs of results (Table 7.2), which come from the construction of the material of the same population as the Y value markings. On the basis of the base regression curve $f(Y)$, theoretical $\hat{Y}$ and Y values, and the differences $\delta Y_i = Y_i - \hat{Y}_i$, the mean value of $\delta Y_{m(n)}$, and standard deviation s should be determined. The parameter of shifting the base curve up to the position corresponding to the lower confidence interval of the corrected curve is determined by:

$$\Delta f = \delta Y_{m(n)} - k_1 s \qquad (7.4)$$

in which the values of the k_1 coefficient are selected from the t-Student distribution, depending on the number of measuring points n. The equation of the scaling curve corrected according to the base standard has the final form:

$$f\left(\hat{Y}\right) = f(Y) + \delta Y_{m(n)} - k_1 s \qquad (7.5)$$

For example, in the case of the compressive strength of concrete, a scaling curve, which was previously adopted according to [7] and is described by the following equation, can be considered as a reliable base regression curve:

$$f_R = 7.4 - 0.915 R + 0.041 R^2 \qquad (7.6)$$

Figure 7.2 shows the correction of the scaling curve described by Equation (7.6).

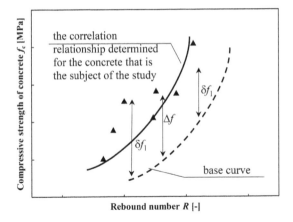

FIGURE 7.2 Correcting the scaling curve described by Equation (7.6) ($\delta f_{1...n}$ – the difference between the unit compressive strength of the concrete core sample and the value of the strength resulting from the base curve).

7.4 CONCLUSIONS

This short chapter highlights the need for calibration and scaling of test equipment used in non-destructive methods. This leads to the correct assessment of the properties of concrete embedded in floors. The problem of fine scaling and rough scaling is highlighted in a synthetic way.

REFERENCES

1. Brunarski, L., & Dohojda, M. (2015). Diagnostics of concrete strength in a structure (in Polish). Warsaw: Building Research Institute.
2. Baboux, J.C., Lakestani, F., Fleischmann, P., & Perdrix, M. (1977). Calibration of ultrasonic transmitters. NDT international, 10(3), 135–138.
3. Bungey, J.H. (1980). The validity of ultrasonic pulse velocity testing of in-place concrete for strength. NDT international, 13(6), 296–300.
4. Popovics, J.S., & Subramaniam, K.V. (2015). Review of ultrasonic wave reflection applied to early-age concrete and cementitious materials. Journal of nondestructive evaluation, 34(1), 267.
5. Wiciak, P., Cascante, G., & Polak, M.A. (2020). Frequency and geometry effects on ultrasonic pulse velocity measurements of concrete specimens. ACI materials journal, 117(2), 205–216.
6. Villain, G., Garnier, V., Sbartaï, Z.M., Derobert, X., & Balayssac, J.P. (2018). Development of a calibration methodology to improve the on-site non-destructive evaluation of concrete durability indicators. Materials and structures, 51(2), 1–14.
7. Runkewicz, L., & Sieczkowski, J. (2018). Assessment of concrete strength in a structure based on sclerometric tests. Guide (in Polish) Warsaw: Building Research Institute.
8. Alwash, M., Breysse, D., Sbartaï, Z.M., Szilágyi, K., & Borosnyói, A. (2017). Factors affecting the reliability of assessing the concrete strength by rebound hammer and cores. Construction and building materials, 140, 354–363.

9. Brencich, A., Bovolenta, R., Ghiggi, V., Pera, D., & Redaelli, P. (2020). Rebound hammer test: An investigation into its reliability in applications on concrete structures. Advances in materials science and engineering, 2020, 6450183.
10. Breysse, D., Romao, X., Alwash, M., Sbartaï, Z.M., & Luprano, V.A. (2020). Risk evaluation on concrete strength assessment with NDT technique and conditional coring approach. Journal of building engineering, 32, 101541.
11. Brencich, A., Cassini, G., Pera, D., & Riotto, G. (2013). Calibration and reliability of the rebound (Schmidt) hammer test. Civil engineering and architecture, 1(3), 66–78.

8 The Diagnosis of Dowelled Concrete Floors

8.1 INTRODUCTION

Dowelled concrete floors are used inside warehouses (Figure 8.1a) and production halls, as well as in hangars and sheds. They are also used in storage and manoeuvring yards, at airports, and on port quays (Figure 8.1b), and in these cases they are called concrete pavements. The implementation of these floors is recommended when the point load, caused, for example, by the movement of heavy means of transport, exceeds 40 kN. Moreover, it is recommended when it is impossible to rule out uneven settling of the expanded areas (slabs) of the floor in relation to each other. As construction practice proves, these floors relatively frequently have significant latent performance flaws, which usually result in the formation of serious failures that have an effect on safe operation.

One of the significant hidden defects in the performance of dowelled concrete floors is their too small thickness in relation to the designed thickness. The thickness of these floors is determined by the designer and is usually in the range from 260 to 400 mm. This defect ultimately translates into a reduced durability when compared to that expected. It is the reason for the breaking of concrete slabs, as well as the breaking of large fragments of corners, which makes it difficult or impossible for the floor to be used safely.

Another important hidden disadvantage is the incorrect dowelling of concrete slabs in strips along expansion joints. This defect may even include several irregularities [1–7]. Defective dowelling makes the cooperation of individual boards with neighbouring ones during the transferring of service loads impossible, does not reduce the bending moments in the slab's corners, and also does not improve the operation of expansion joints, i.e., by causing their smaller expansion. This can lead to the keying of the dilated floor panels (fields), the cracking and breaking off of corners, and also a reduction of the projected durability and safety of use.

These disadvantages often add up, in turn increasing the problem of reduced durability and danger during use.

8.2 REASONS FOR HIDDEN DEFECTS OF DOWELLED CONCRETE FLOORS

As mentioned in Section 8.1, the significant hidden disadvantages of dowelled concrete floors, which reduce their durability and safety of use, include insufficient thickness and the incorrect dowelling of concrete slabs. The risk of these defects

DOI: 10.1201/9781003288374-8

FIGURE 8.1   Examples of dowelled concrete floors.

arises at the execution stage in the absence of an effective and systematic control of the works in the form of technical supervision.

As evidenced by construction practice, the reason for too low thickness in relation to the designed thickness is the incorrect (often only rough) levelling of the upper surface (without the use of geodetic methods) of the substrate under the floor. In a situation where the desired level of the upper floor surface is geodetic while laying the concrete mix, the thickness of the floor is not the same over the entire surface. It is obvious that areas that are too thin are undesirable. An example of this can be seen in the test results shown in Figure 8.2, which show that the thickness of the floor, which should be 300 mm, is in the range between 230–295 and 259.87 mm on average.

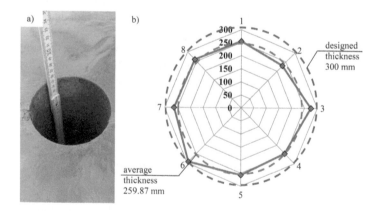

FIGURE 8.2   Testing the thickness of an exemplary dowelled concrete floor: (a) measurement of the thickness in one of the eight core boreholes made, (b) the results of the measurement of the thickness in the floor.

In an extreme case, the lack of thickness is over 23%, and such places are the first to be damaged during operation.

Faulty dowelling of concrete slabs can be caused by numerous irregularities:

- the chaotic arrangement of individual steel dowels – not parallel to each other, at different intervals, not in the lines of cut expansion slits, and not perpendicular to the plane of these slots,
- not horizontal arrangement of steel dowels, with them not being in the middle of the slab's thickness,
- displacement of the composite structure made of steel dowels that are connected with a wire, which is caused by the pressure of the laid and vibrated concrete mix,
- cutting expansion joints in bands that are different to those designed,
- no steel dowels at all.

The placement of steel dowels is illustrated in Figure 8.3. It is clear that the existence of the above mentioned irregularities proves that the contractor and supervisor lack awareness concerning the very important role of steel dowels. This role is especially important when ensuring the proper support of individual dilated floor fields.

It still requires clarification that dowels should be made of smooth steel bars without hooks. In practice, their diameter is usually from 16 to 32 mm, and their length is from 500 to 700 mm, with their spacing most often being from 300 to 400 mm. They can be made of single bars (Figure 8.3), but structures (so-called "baskets") are also used, which consist of steel bars connected with a wire and placed on steel supports (Figure 8.4).

8.3   METHODOLOGY OF THE NON-DESTRUCTIVE DIAGNOSIS OF DOWELLED CONCRETE FLOORS

Figure 8.5 presents the authors' original methodology, of the non-destructive diagnosis of dowelled concrete floors in terms of the non-destructive detection of the invisible defects described in Section 8.2, i.e., too small thickness in relation to that designed, as well as defective dowelling. This methodology is based on the use of the ultrasonic tomography method. It can be useful in both tests of newly constructed floors prior to their commissioning, as well as in diagnosing exploited floors. For a better understanding, this methodology is illustrated with exemplary results obtained for a doweled concrete floor tested by the authors.

The first step in the developed methodology is to test the thickness of the concrete floor using a non-destructive ultrasonic tomography method (Figure 8.6). In this step, start by positioning the ultrasound tomograph antenna on the top surface of each dilated floor area at randomly selected 500 × 100 mm test sites, and then perform the tests. Then record the results and prepare flat images in three mutually perpendicular directions. Based on a detailed analysis of these images, the thickness of the floor should be determined at each test site. In the case of the ultrasonic tomography method, the accuracy of this determination is 10 mm. The results obtained in a non-destructive way by the ultrasonic tomography method need to be validated with control boreholes, for example, by a control section in randomly selected places.

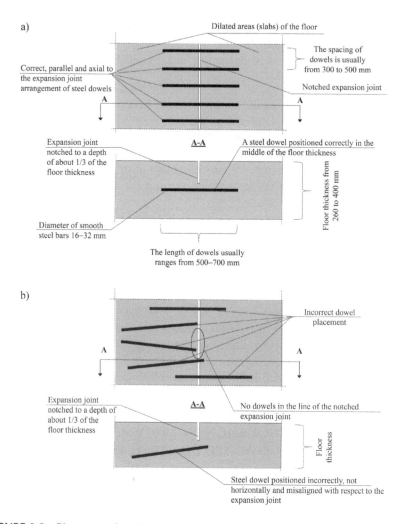

FIGURE 8.3   Placement of steel dowels in a concrete floor: (a) correct, (b) incorrect.

FIGURE 8.4   View of steel bars connected with a wire (the so-called "basket").

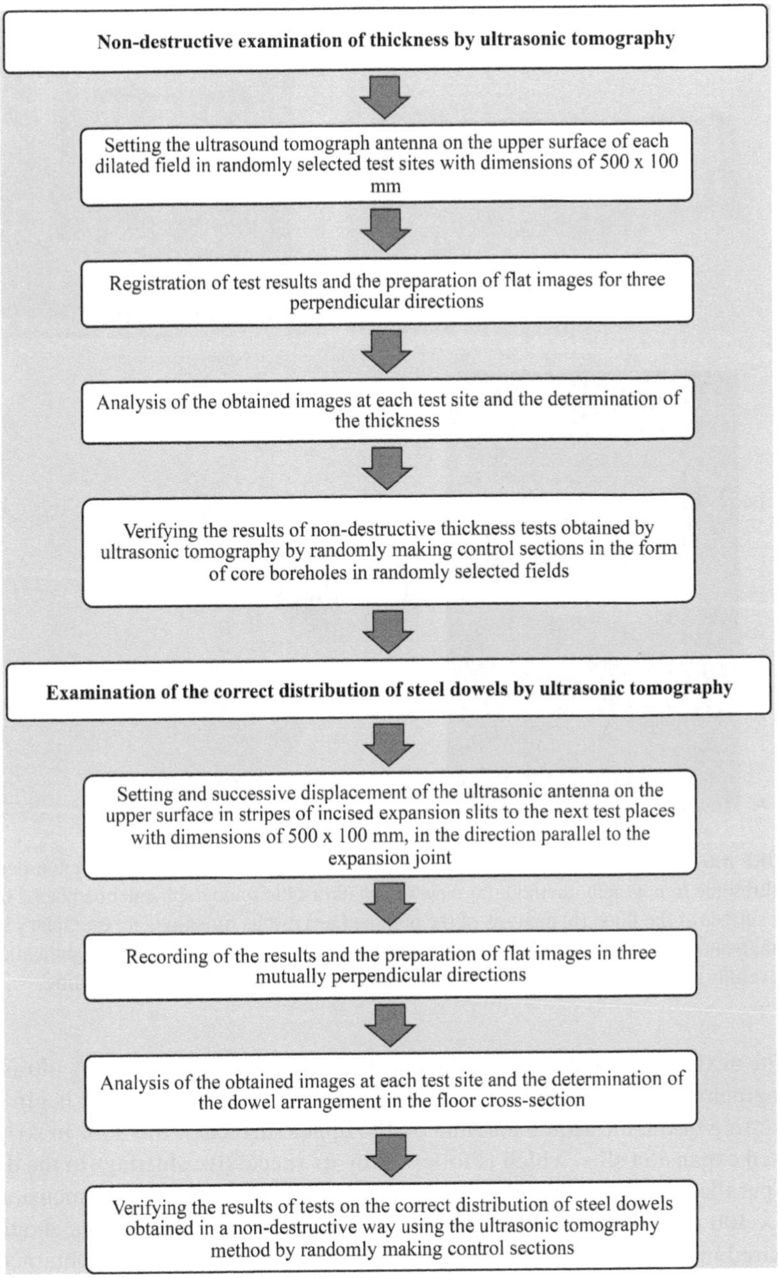

Non-destructive examination of thickness by ultrasonic tomography

Setting the ultrasound tomograph antenna on the upper surface of each dilated field in randomly selected test sites with dimensions of 500 x 100 mm

Registration of test results and the preparation of flat images for three perpendicular directions

Analysis of the obtained images at each test site and the determination of the thickness

Verifying the results of non-destructive thickness tests obtained by ultrasonic tomography by randomly making control sections in the form of core boreholes in randomly selected fields

Examination of the correct distribution of steel dowels by ultrasonic tomography

Setting and successive displacement of the ultrasonic antenna on the upper surface in stripes of incised expansion slits to the next test places with dimensions of 500 x 100 mm, in the direction parallel to the expansion joint

Recording of the results and the preparation of flat images in three mutually perpendicular directions

Analysis of the obtained images at each test site and the determination of the dowel arrangement in the floor cross-section

Verifying the results of tests on the correct distribution of steel dowels obtained in a non-destructive way using the ultrasonic tomography method by randomly making control sections

FIGURE 8.5   Methodology of the non-destructive diagnosis of dowelled concrete floors.

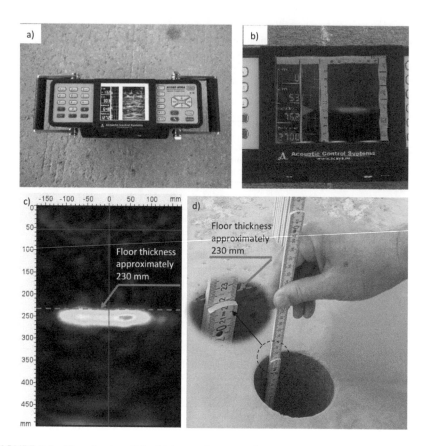

FIGURE 8.6 Examination of the thickness of an exemplary concrete floor with a non-destructive ultrasonic tomography method: (a) view of the ultrasonic tomograph antenna placed on the upper surface of the floor, (b) analysis of the obtained test results (imaging), (c) exemplary image of a fragment of the floor with a diagnosed thickness of approximately 230 mm, (d) authentication of the results of non-destructive tests using a control section in the form of core drilling.

The next step is to investigate the distribution of steel dowels using ultrasonic tomography, as illustrated in Figures 8.7 and 8.8. The tests should begin with the setting of the ultrasonic antenna on the upper surface of the floor in strips of incised expansion slits, which is followed by its successive shifting (in the direction parallel to the expansion joint) to the next test sites that have dimensions of 500 × 100 mm. The results should then be recorded and flat images should be prepared in three perpendicular directions. Detailed analysis of the obtained flat images at each test site enables the assessment of the dowel arrangement in the floor with an accuracy of 10 mm. In situations where there is doubt, it is possible to build a three-dimensional image from flat images, which can then be analysed.

It can be seen from Figure 8.8 that the arrangement of the steel dowels in the cross-section of the floor, which is determined in the tested example floor by means of ultrasonic tomography, is defective. The spacing of the dowels ranges from 27 to 42 cm –

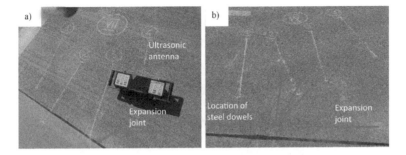

FIGURE 8.7 Non-destructive testing of the distribution of steel dowels in the cross-section of an exemplary floor: (a) searching for the location of dowels along the expansion joint cut in the floor using the ultrasonic tomography method, (b) an exemplary test result in the form of a localized position of dowels that are shifted significantly beyond the strip with an incised expansion joint.

in each case the dowels are shifted in relation to the axis of the expansion joint, with one dowel (No. 4) running diagonally. Figure 8.8 also shows that the depth of the dowels in the vicinity of the expansion joint is in the range of 13–22 cm, and at the end of the dowel it is in the range of 16–24 cm. As can be seen from Figure 8.8, all the dowels are not placed in the middle of the thickness of the concrete floor, with seven of them not being placed horizontally.

Moreover, the test results shown in Figure 8.8 document the exemplary views of the fragments of the tested concrete floor presented in Figure 8.9. The fragments include the area near the expansion joint, and also the area at the end of dowels 3, 5, and 10.

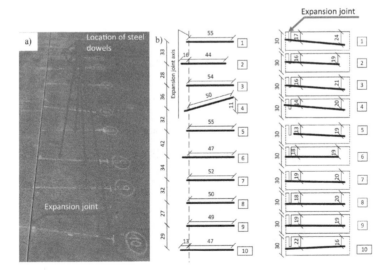

FIGURE 8.8 View of: (a) the upper surface of the tested concrete floor with the locations of steel dowels marked on it – localized using the ultrasonic tomography method, (b) a sketch of the dowel arrangement in the floor in the top view and a sketch of the dowel arrangement in the cross-section of the concrete floor.

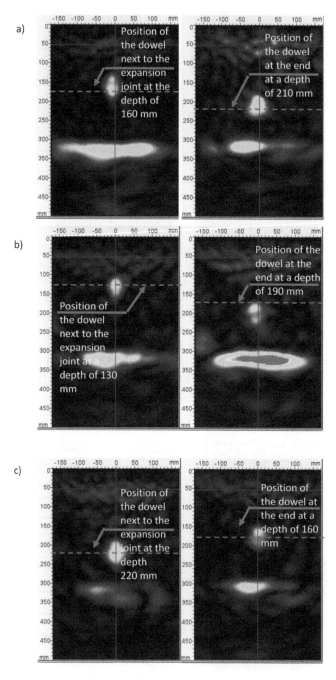

FIGURE 8.9   View of the images obtained with an ultrasonic tomograph on a fragment of the concrete floor near the expansion joint and at the end of the steel dowels: (a) no. 3, (b) no. 5, (c) no. 10.

The last step of the developed test methodology is to validate the results of the tests of the correct distribution of the steel dowels in the cross-section of the concrete floor, which were obtained non-destructively using the ultrasonic tomography method. This validation is conducted by randomly making control exposures.

8.4   CONCLUSIONS

In dowelled concrete floors, serious damage occurs relatively often, preventing safe use and reducing durability. They are the result of significant hidden defects arising at the execution stage. In this chapter, these disadvantages have been indicated, discussed and supported by exemplary results from the author's own experience. This chapter also includes the developed original methodology of non-destructive performance of these defects, enriched with credible examples of results from own research. This methodology can be useful both in diagnosing dowelled concrete floors already in use as well as in tests of newly made floors during their quality acceptance before putting into use.

REFERENCES

1. Jung, Y.S., & Zollinger, D.G. (2011). New laboratory-based mechanistic–empirical model for faulting in jointed concrete pavement. Transportation research record, 2226(1), 60–70.
2. Khazanovich, L., Darter, M.I., & Yu, H.T. (2004). Mechanistic-empirical model to predict transverse joint faulting. Transportation research record, 1896(1), 34–45.
3. Bakhsh, K.N., & Zollinger, D. (2014). Faulting prediction model for design of concrete pavement structures. In Pavement materials, structures, and performance (pp. 327–342).
4. Seruga, A., Seruga, T., & Juliszewski, L. (2011). Dowels in concrete road surfaces (in Polish). Czasopismo Techniczne. Budownictwo, 108, 113–134.
5. Maitra, S.R., Reddy, K.S., & Ramachandra, L.S. (2009). Load transfer characteristics of dowel bar system in jointed concrete pavement. Journal of transportation engineering, 135(11), 813–821.
6. Szydło, A. (2004). Cement concrete road pavements (in Polish). Kraków: Polski Cement.
7. Choi, S., Ha, S., & Won, M.C. (2011). Horizontal cracking of continuously reinforced concrete pavement under environmental loadings. Construction and building materials, 25(11), 4250–4262.

9 The Diagnostics of Stone Floors

9.1 INTRODUCTION

Concrete floors in large representative commercial and public facilities (Figure 9.1) are usually finished with stone slabs (most often marble or granite) that are laid on a cement substrate [1–4]. The dimensions of the slabs can be large, even up to 1000 × 1000 mm, with their thickness being from 30 to 50 mm. The technology of making floors from such slabs seems to be simple. The slabs are placed on the substrate, which is a layer of cement mortar with a moist consistency, and the substrate is located on a concrete substrate, which is, for example, a ceiling slab. Immediately after laying the mortar and it being levelled, cement laitance is gradually and evenly spread on the surface of the fresh substrate, constituting the bonding layer. On the prepared substrate, stone slabs are immediately laid and levelled.

Figure 9.2 summarizes the most important technological and technical requirements for the implementation of concrete floors with a top layer made of stone slabs.

In construction practice, there are cases where newly made concrete floors with a top layer made of stone slabs are seriously damaged after short-term use. This damage manifests itself in the form of:

- cracks (breaking) of plates,
- chipping off of corners, and
- chipping edges.

These damages can be considered typical and are usually caused by:

- local deficiencies of the cement mortar that constitutes the substrate under the stone slabs and its lack of levelling,
- not spreading the cement laitance over the entire surface of the cementitious substrate, and
- too small thickness in relation to the designed thickness of the stone slabs.

Cracked stone slabs and slabs with broken corners and jagged edges not only reduce the aesthetic value, but also the durability and safety of using the floor.

9.2 DESCRIPTION OF CAUSES OF DAMAGE TO CONCRETE FLOORS THAT HAVE THE TOP LAYER MADE OF STONE SLABS

When diagnosing damaged concrete floors that have the top layer made of stone slabs, local deficiencies in the cement mortar under the stone slabs, and their lack of alignment, are very often found. This irregularity results in cracking of the stone

DOI: 10.1201/9781003288374-9

FIGURE 9.1   Sample views of the surface of concrete floors that have the top layer made of stone slabs.

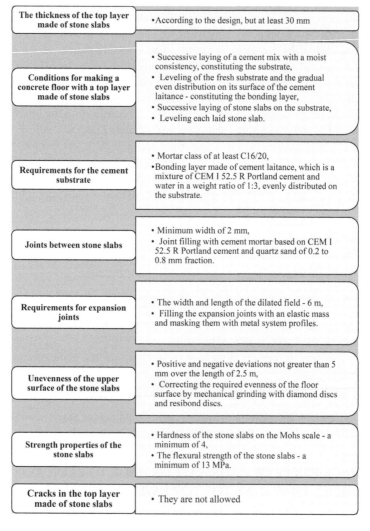

The thickness of the top layer made of stone slabs	• According to the design, but at least 30 mm
Conditions for making a concrete floor with a top layer made of stone slabs	• Successive laying of a cement mix with a moist consistency, constituting the substrate, • Leveling of the fresh substrate and the gradual even distribution on its surface of the cement laitance - constituting the bonding layer, • Successive laying of stone slabs on the substrate, • Leveling each laid stone slab.
Requirements for the cement substrate	• Mortar class of at least C16/20, • Bonding layer made of cement laitance, which is a mixture of CEM I 52.5 R Portland cement and water in a weight ratio of 1:3, evenly distributed on the substrate.
Joints between stone slabs	• Minimum width of 2 mm, • Joint filling with cement mortar based on CEM I 52.5 R Portland cement and quartz sand of 0.2 to 0.8 mm fraction.
Requirements for expansion joints	• The width and length of the dilated field - 6 m, • Filling the expansion joints with an elastic mass and masking them with metal system profiles.
Unevenness of the upper surface of the stone slabs	• Positive and negative deviations not greater than 5 mm over the length of 2.5 m, • Correcting the required evenness of the floor surface by mechanical grinding with diamond discs and resibond discs.
Strength properties of the stone slabs	• Hardness of the stone slabs on the Mohs scale - a minimum of 4, • The flexural strength of the stone slabs - a minimum of 13 MPa.
Cracks in the top layer made of stone slabs	• They are not allowed

FIGURE 9.2   Summary of the most important requirements for the implementation of concrete floors that have the top layer made of stone slabs.

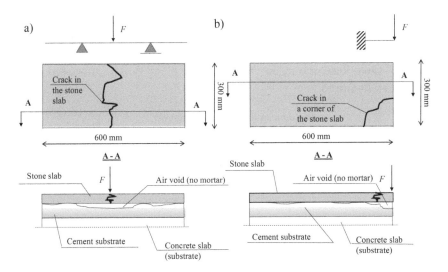

FIGURE 9.3 The mechanism of cracking of stone slabs during operation as a result of: (a) a local lack of cement mortar under the slab and its lack of alignment resulting in a "point" support of the slab, (b) a lack of cement mortar under the slab in the corner area, resulting in a cantilever support of the slab corners.

slabs due to their "point" support during operation, as well as due to a "point" load caused by, for example, a wheel of a freight trolley. The cracking mechanism is explained in Figure 9.3(a). When there is no mortar in the area of the slab corner, the loaded corner becomes a support, which then breaks off during operation. The mechanism of this damage is explained in Figure 9.3(b). It is important that this mechanical damage is not repairable. A troublesome replacement of such damaged boards with new ones is required, because, as was stated in Figure 9.2, cracks in the stone slabs are not allowed.

For example, Figure 9.4 shows a view of typical damage (formed during the use of the floor) in the form of a stone slab fracture in the middle of its span due to its

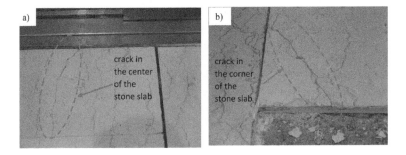

FIGURE 9.4 View of typical cracks formed in the exploited concrete floor that has the top layer made of stone slabs: (a) in the middle of the span of the stone slab due to its "point" support, (b) in the corner of the stone slab due to its cantilever support.

TABLE 9.1

Examples of the Average Values of Strength Parameters Obtained in the Authors' Own Research [5] for Marble Stone Slabs When Compared to the Values Required

The Tested Strength Parameter of the Stone Slab	The value of the Strength Parameter Obtained on the Basis of Tests		The Required Value of the Strength Parameter Along and Across the Split Line
	Along the Split Line	Across the Split Line	
Flexural strength	12.2 MPa	12.2 MPa	13.8 MPa

"point" support, and a corner of the stone slab breaking off due to its cantilever support.

These cracks may also be the result of the accumulation of irregularities in the preparation of the cement substrate that has a lower flexural strength than the required values of the flexural strength of the stone slabs. Examples of average values of the flexural strength of stone slabs obtained on the basis of own research [5] for a floor made of marble slabs, in comparison with the requirements, are summarized in Table 9.1. The flexural strength was determined in accordance with the standard [6]. Twenty-four samples were used for the tests, 12 of which were cut along the split line (flexural load perpendicular to the anisotropy plane), with the remaining 12 samples being cut across the split line (flexural load parallel to the anisotropy plane).

When commenting on the results in Table 9.1, it should be noted that the average value of the flexural strength of the tested slab samples in the bending strength test is 12.2 MPa both along and across the interfacing line, which is lower by more than 13% than the required value. This means that due to their too low flexural strength, they are more susceptible to mechanical damage caused by point loads.

Chipping of the edges of individual stone slabs is also a problem that is troublesome to remove. An example of such a typical failure is shown in Figure 9.5. These chipped edges usually arise as a result of the passage of various types of transport, both during construction and use. The cause of their formation is the same as in the case of the breaking off of the corners (Figure 9.3). Chipped plates must be replaced with new ones, in the same way as cracked slabs or slabs with broken corners must also be replaced.

In the case of concrete floors that have the top layer made of stone slabs, a very frequent technological irregularity is the lack of spreading of the cement laitance over the entire surface of the cement substrate. The importance of this treatment for the quality and durability of these floors is not appreciated by contractors or technical supervision. The cement laitance is a bonding layer and determines the proper bonding of stone slabs with the cement substrate. If this combination is absent, delamination forms in its place. This promotes cracking of the plates during operation, meaning that the required durability and safety of the floor in use

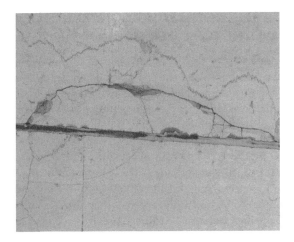

FIGURE 9.5 View of a chipped edge of a stone slab [5].

will not be sufficient. For this reason, before quality acceptance and the commissioning of floors that have stone slabs, this combination should be unconditionally assessed.

Another observed problem with these floors is sometimes the thickness of the stone slabs being too small in relation to the designed thickness. For example, the results presented in Figure 9.6 show (for a floor investigated by the authors [5]) that the thickness of the stone slabs, which should be 30 mm, is too small. It is in the range between 19.8 and 20.5 mm, and amounts to an average of 20.11 mm, which means a thickness deficiency of almost 30%. Embedding into the surface layer of the concrete floor with such a large shortage of thickness should not take place.

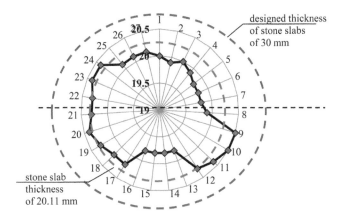

FIGURE 9.6 The results of tests of the thickness of stone slabs embedded in the floor.

9.3   METHODOLOGY OF TESTING CONCRETE FLOORS THAT HAVE THE TOP LAYER MADE OF STONE SLABS

An effective counteraction to the occurrence of the above-mentioned damage to stone slabs that constitute the top layer of concrete floors is to prevent the causes of their formation. The original two-stage methodology of control and research, which was developed by the authors, may be useful for this purpose. This original methodology for testing concrete floors that have the top layer made of stone slabs is shown graphically in Figures 9.7–9.9, and is also described below. Figure 9.7 shows an overview, and Figures 9.8 and 9.9 show a detailed diagram of the methodology. The first stage of this methodology is dedicated to the execution stage of the floor, while the second stage is useful when testing before the technical acceptance that allows floors to be used and to be diagnosed during operation. At this stage, it is recommended to use non-destructive research methods.

Stage 1 involves checking the implementation of the floors that have the top layer made of stone slabs (Figure 9.8). The first step in this stage is to check that the thickness of the stone slabs meets the design requirements. This step may be combined with testing the flexural strength and abrasion of the stone slabs. The next step is to control the consistency of the cement mixture that is used to make the substrate for the stone slabs. In accordance with the requirements given in Figure 9.2, this consistency should be moist. Then, using the visual method and with the support of a non-destructive geodetic method, it is necessary to control the thickness, evenness and level of the laid layer of cement mortar. The next step is to control the implementation of the bonding layer of cement laitance using a direct visual method, and then to control the levelling of the laid boards using a non-destructive geodetic method. The unevenness of the upper surface of the floor is only permissible while maintaining the upper and lower deviation. The next step is to control the execution of expansion joints in terms of maintaining the appropriate size of the dilated field,

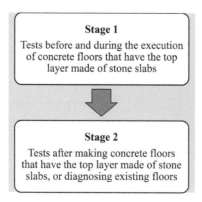

FIGURE 9.7   General diagram illustrating the developed methodology of testing concrete floors that have the top layer made of stone slabs.

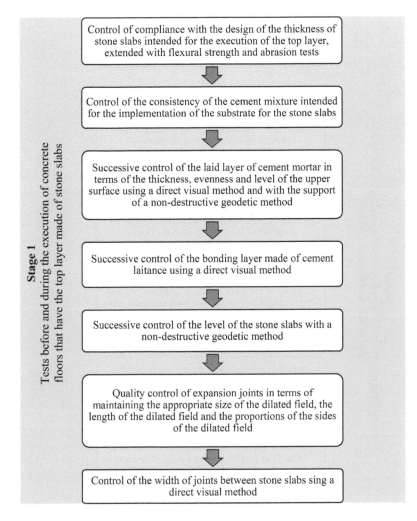

FIGURE 9.8 Methodology of performance control and tests at the execution stage of concrete floors that have the top layer made of stone slabs – stage 1.

the length of the dilated field, and the proportions of the sides of the dilated field. Then, the expansion joint widths should be tested using the direct visual method, followed by filling them with a permanently elastic mass and masking them with metal system profiles.

Stage 2 includes tests after the floor has been made and before technical acceptance that allows it to be used, as well tests of already operated and diagnosed floors (Figure 9.9). The first step in stage 2 is a rough quality control of the bonding between the stone slabs and the cement substrate. Such control can be executed using the tapping method, for example, with a wooden element moving along one direction, and in selected areas that raise doubts. In places where

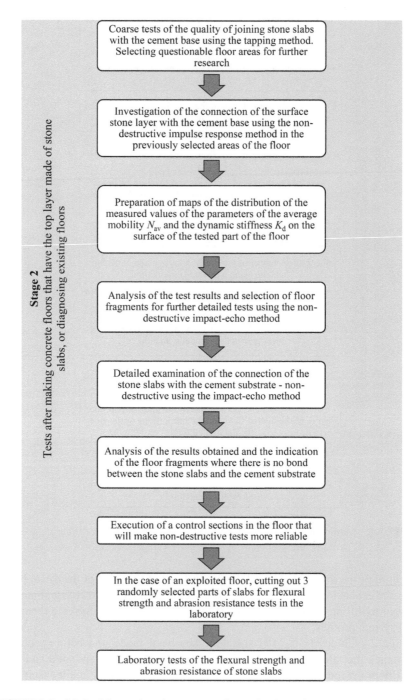

FIGURE 9.9 Methodology of testing concrete floors that have the top layer made of stone slabs – stage 2.

a)

b)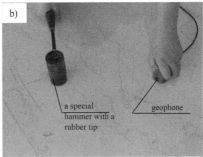

FIGURE 9.10 Non-destructive investigation of the bonding of the stone slabs with the cement substrate using the impulse response method: (a) conducting research, (b) close-up of the research site.

noise is heard, it should be suspected that the cement mortar is not bonding with the stone slabs.

The questionable areas (fields) should then be subjected to non-destructive testing using impulse response and impact-echo methods. First, the impulse response method should be used and a grid of measurement points at a distance of 1000 mm should be applied to this area of the floor. At individual test points, the elastic wave should be excited with a special hammer with a rubber tip. The elastic wave should be recorded with a geophone and the values of the parameters of the average mobility N_{av} and the dynamic stiffness K_d should be determined (Figure 9.10). On the basis of the obtained results, maps of the distribution of the values of these parameters on the surface of the tested part of the floor should be prepared (Figure 9.11).

The prepared maps should then be analysed, and fragments of the floor in which the parameter of the average mobility N_{av} and dynamic stiffness K_d have a low value should be selected. The local increase in the value of the average mobility N_{av} parameter on the analysed map, and the accompanying low value of the dynamic stiffness K_d parameter mean that the floor is susceptible to "sagging", which may indicate a lack of bonding between the layers and the presence of delamination at the interface between the cement substrate and the stone slabs.

Then, the selected fragments should be tested using the non-destructive impact-echo method (Figure 9.12). These tests should be carried out with the mesh of points densified to 100 mm. Examples of test results in one of such points are illustrated in Figure 9.13. In the amplitude-frequency spectra shown in this figure, one can distinguish the dominant frequency f_T corresponding to the thickness of the element (total thickness of the concrete slab, cement base and stone slab), and also the frequency f_D corresponding to the reflection of the elastic wave from the delamination between the stone slabs and the cement substrate. On the basis of these tests, it is necessary to indicate those fragments where there is no bonding of the stone slabs with the cement substrate.

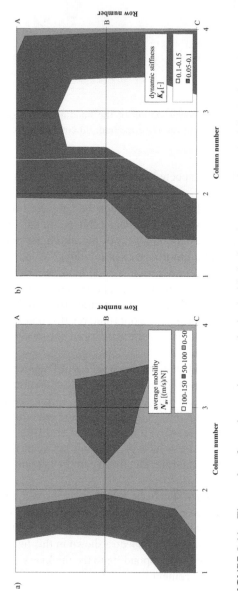

FIGURE 9.11 The results of non-destructive tests obtained with the impulse response method, in the form of parameter distribution maps: (a) average mobility N_{av}, (b) dynamic stiffness K_d.

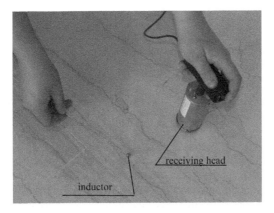

FIGURE 9.12 Non-destructive tests of the bonding of the stone slabs with the cement substrate using the non-destructive impact-echo method.

In order to validate the non-destructive test results, a control section should be made by disassembling one of the boards from the floor, as shown in Figure 9.14.

If a floor in use is the subject of the research, then as part of its diagnostics, flexural strength tests of its stone slabs should be performed on the basis of randomly

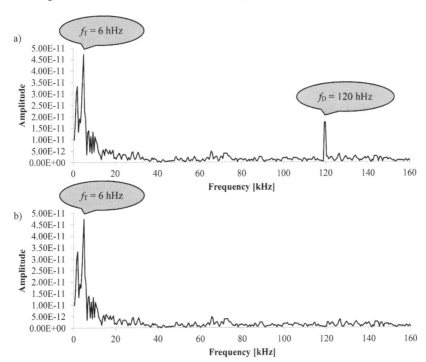

FIGURE 9.13 Examples of the amplitude-frequency spectra of an elastic wave recorded using the non-destructive impact-echo method: (a) indicating the presence of delamination at the interface between the stone slab and the substrate, (b) in the absence of delamination.

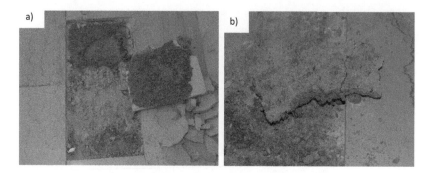

FIGURE 9.14 View of the control section confirming a lack of bonding, which was investigated on the basis of non-destructive tests: (a) unevenness of the cement substrate and a lack of cement laitance, (b) strong sporulation of the substrate mortar.

taken samples from the floor. According to [6], the flexural strength should first be determined. For this purpose, three fragments of slabs (possibly three whole slabs) should be taken for testing. Afterwards, in order to conduct flexural strength tests, at least three samples should be cut from the slabs: 120–150 mm long, 50–60 mm wide, and with a thickness equal to the thickness of the floor. The increase of the destructive force during the flexural strength tests should amount to 0.25 MPa/s. From the taken fragments of the floor, at least three samples – with a length of 71 mm, a width of 71 mm, and a height equal to the thickness of the top layer of the floor – should also be prepared in order to test their abrasion resistance according to [7].

9.4 CONCLUSIONS

This chapter, devoted to the diagnosis of concrete floors that have the top layer made of stone slabs, illustrates the damage that stone slabs often suffer after a short-term use. The causes and mechanisms of these damages were explained, and examples of own research were presented. In order to counteract the occurrence of damage, the authors proposed a methodology for checking the execution of these floors and testing of newly-made floors as part of technical acceptance before they are allowed to use. This methodology is also useful in diagnosing exploited floors.

REFERENCES

1. Nicoletti, G.M., Notarnicola, B., & Tassielli, G. (2002). Comparative life cycle assessment of flooring materials: Ceramic versus marble tiles. Journal of cleaner production, 10(3), 283–296.
2. Springfield, J. (2004). Use of clay tiles in floor and wall construction. Practice periodical on structural design and construction, 9(1), 2–4.
3. Carrino, L., Polini, W., & Turchetta, S. (2002). An automatic visual system for marble tile classification. Proceedings of the institution of mechanical engineers, part b. Journal of engineering manufacture, 216(8), 1095–1108.

4. Stainton, S. (1980). The care of floors and floor coverings. Studies in conservation, 25(1), 44–46.
5. Hoła, J., Sadowski, Ł., & Nowacki, A. (2019). Analysis of the causes of cracks in marble slabs in a large-surface floor of a representative commercial facility. Engineering failure analysis, 97, 1–9.
6. EN 12372:2007, Natural stone test methods – Determination of flexural strength under concentrated load.
7. EN 14157:2017, Natural stone test methods – Determination of the abrasion resistance

10 The Diagnostics of Concrete Industrial Floors

10.1 INTRODUCTION

Currently, industrial floors are made in the form of a concrete or reinforced concrete slab, with the upper surface finished either by rubbing or with the use of various types of toppings [1–3]. Such floors are widely used in large-scale industrial and warehouse facilities, of which more and more have recently been built. Concrete reinforced with steel fibres, and reinforcement bars, is used to make these floors. Diffuse reinforcement and rebars are also used [4–7]. There is no doubt that the entire process of making concrete industrial floors should comply with the conditions and requirements specified by the designer – there is then a guarantee that the concrete industrial floor will meet the requirements set for it [8]. The most important of these requirements is given in Figure 10.1.

In construction practice, there are numerous cases of industrial concrete floors that are defective and significantly damaged after short-term use. This lowers their durability and safety of use, leading to their shutdown. When diagnosing these floors, the following can often be found:

- too small thickness in relation to the designed thickness, which is an invisible defect,
- the occurrence of unacceptable cracks,
- the presence of unacceptable hairline and mesh cracks on the surface,
- too low strength parameters in relation to those required.

According to the authors, the formation of these defects is mainly due to the lack of methodical control of the floors during their execution, which means that the requirements, listed in Figure 10.1, are not met.

10.2 EXAMPLES OF THE MOST COMMON DEFECTS IN INDUSTRIAL CONCRETE FLOORS

As mentioned in Section 10.1, the thickness of an industrial concrete floor is often found to be too small in relation to its designed thickness. For example, Figure 10.3 shows example of research results that prove the existence of this kind of defect in one of the floors investigated by the authors Figure 10.2. As shown in Figure 10.3, its thickness, which should be 180 mm, falls within the range between 139 and 193 mm,

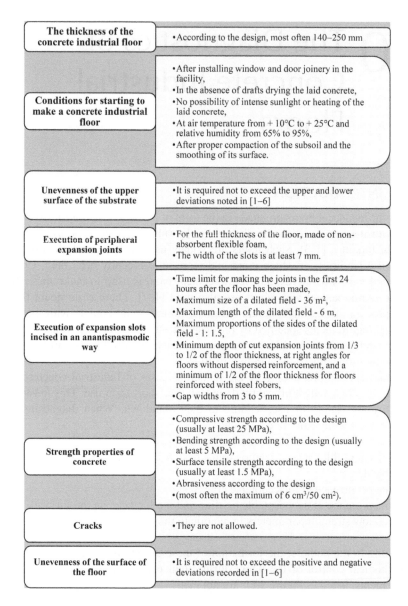

The thickness of the concrete industrial floor	• According to the design, most often 140–250 mm
Conditions for starting to make a concrete industrial floor	• After installing window and door joinery in the facility, • In the absence of drafts drying the laid concrete, • No possibility of intense sunlight or heating of the laid concrete, • At air temperature from + 10°C to + 25°C and relative humidity from 65% to 95%, • After proper compaction of the subsoil and the smoothing of its surface.
Unevenness of the upper surface of the substrate	• It is required not to exceed the upper and lower deviations noted in [1–6]
Execution of peripheral expansion joints	• For the full thickness of the floor, made of non-absorbent flexible foam, • The width of the slots is at least 7 mm.
Execution of expansion slots incised in an anantispasmodic way	• Time limit for making the joints in the first 24 hours after the floor has been made, • Maximum size of a dilated field - 36 m², • Maximum length of the dilated field - 6 m, • Maximum proportions of the sides of the dilated field - 1: 1.5, • Minimum depth of cut expansion joints from 1/3 to 1/2 of the floor thickness, at right angles for floors without dispersed reinforcement, and a minimum of 1/2 of the floor thickness for floors reinforced with steel fobers, • Gap widths from 3 to 5 mm.
Strength properties of concrete	• Compressive strength according to the design (usually at least 25 MPa), • Bending strength according to the design (usually at least 5 MPa), • Surface tensile strength according to the design (usually at least 1.5 MPa), • Abrasiveness according to the design • (most often the maximum of 6 cm³/50 cm²).
Cracks	• They are not allowed.
Unevenness of the surface of the floor	• It is required not to exceed the positive and negative deviations recorded in [1–6]

FIGURE 10.1 Summary of the most important requirements for the implementation of concrete industrial floors.

depending on the test site, and on average amounts to 161.62 mm. This means that, in extreme cases, the thickness deficiency is significant and amounts to about 23%, with the average thickness deficiency being over 11%.

According to Figure 10.1, the occurrence of cracks in an industrial concrete floor is not allowed. This defect may be the result of, inter alia, the incorrect shape of expansion joints made, for example, of the columns from the floor. For example,

FIGURE 10.2   General view of an industrial concrete floor tested by the authors.

Figure 10.4 shows how the correct expansion should be performed, and also how the incorrect expansion was performed.

Cracks are also the result of the improper execution of cut expansion joints, i.e., making the joints too late, too large expansion fields, too long expansion fields, exceeding the recommended maximum proportions of the sides of the expanded field, a too small depth of the cut expansion joints. The expansion joints made in this way do not play an antispasmodic role. Examples of these abnormalities are illustrated in Figure 10.5.

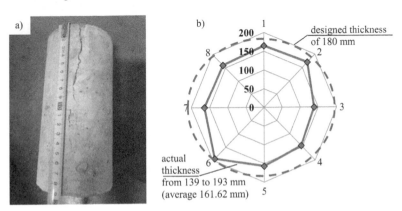

FIGURE 10.3   The results of measuring the thickness of an industrial concrete floor: (a) view of one of the core samples taken from the floor for testing, (b) the results of the measurement.

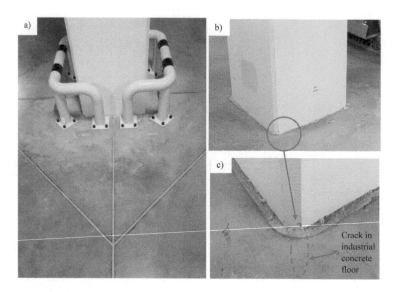

FIGURE 10.4   Separation of a column in a concrete industrial floor: (a) correct, in the form of A cut slit ("diamond" form), (b) incorrect, causing the floor to crack, (c) close-up of the resulting crack.

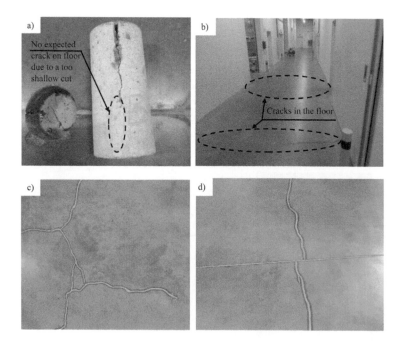

FIGURE 10.5   Effects of the improper incision of antispasmodic expansion slots in the concrete floor: (a) too shallow depth of the cut expansion joint, (b) cracks due to exceeding the maximum proportions of the sides of the dilatation field, (c) multi-directional cracks caused by the lack of cut slots, (d) crack perpendicular to the cut expansion joint caused by exceeding the size of the dilated field.

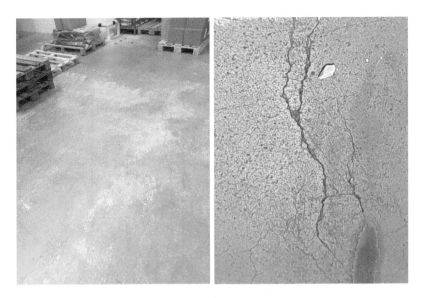

FIGURE 10.6   General and close-up view of unacceptable local abrasions on the surface.

Figures 10.6 and 10.7, on the other hand, show an exemplary view of unacceptable local abrasions and hairline and mesh cracks in concrete floors, which result in generally reduced near-surface tensile strength and increased abrasion.

By diagnosing industrial concrete floors, it can be concluded that they are very often characterized by a too low compressive strength of concrete. This statement is related especially to the upper zone of the floor. Based on [9], Figure 10.8 shows

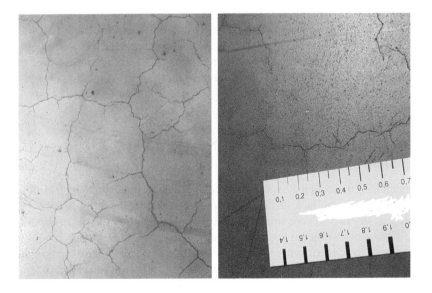

FIGURE 10.7   View of unacceptable hairline and mesh scratches on the surface.

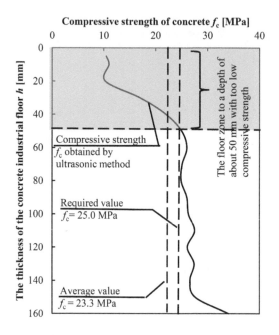

FIGURE 10.8   A test result of the compressive strength of concrete along the thickness of a concrete industrial floor determined by the non-destructive ultrasonic method.

exemplary test results of the course of the compressive strength of concrete along the thickness of an industrial concrete floor obtained with the use of the non-destructive ultrasonic method. In Figure 10.8, the required average compressive strength of 25 MPa and the average strength of 23.3 MPa obtained from the tests are marked with vertical lines. Moreover, the non-vertical continuous line shows the compressive strength determined on the basis of non-destructive ultrasonic tests, and also the developed correlation relationship – as mentioned earlier in Chapter 3. It shows that up to a depth of about 50 mm (counting from the top surface of the industrial concrete floor), the compressive strength of the concrete is lower than the required strength. This strength is drastically too low in the 20 mm-thick upper zone of the floor.

10.3   METHODOLOGY OF TESTING CONCRETE INDUSTRIAL FLOORS

As mentioned in Section 10.1, the formation of defects in concrete industrial floors is mainly due to the lack of methodological and effective control of their implementation. In order to counteract this, the authors (based on their own experience) have developed an original three-stage methodology, which is not only useful when testing prior to the commencement of constructing a concrete industrial floor newly made floors, during the construction of a concrete industrial floor, before handing them over for use, but also when diagnosing already used floors. This methodology was developed while taking into account useful non-destructive testing methods.

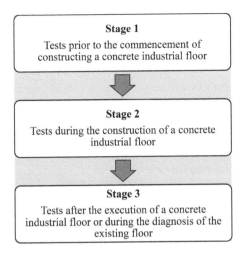

FIGURE 10.9 General scheme of the methodology of testing concrete industrial floors.

The developed original methodology is shown graphically in Figures 10.9–10.11, and described below. Figure 10.9 shows an overview, and Figures 10.10 and 10.11 show a detailed diagram of the methodology.

Stage 1 consists of tests prior to commencing the production of an industrial concrete floor (Figure 10.10). In this stage, it is necessary to check whether the building has built-in window and door joinery. Building in the woodwork is necessary because it prevents the formation of so-called drafts, which result in rapid evaporation of water from the concrete mixture. Next, the temperature and relative air humidity in the facility should be checked, which, according to the requirements given in Figure 10.1, should

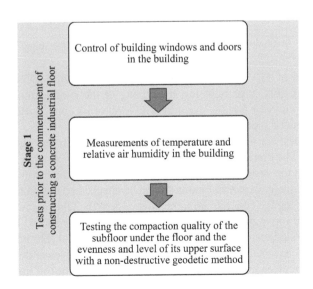

FIGURE 10.10 Methodology of testing concrete industrial floors – stage 1.

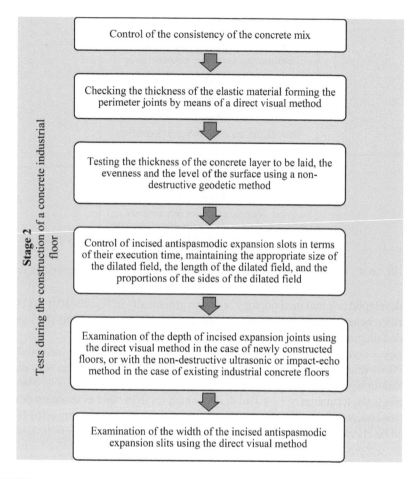

FIGURE 10.11 Methodology of testing concrete industrial floors – stage 2.

be in the range of +10°C to +25°C for temperature, and 65%–95% for humidity. The last step in this stage is to test the testing of the compaction quality of the subfloor under the floor and the evenness and level of its upper surface with a non-destructive geodetic method.

Stage 2 consists of tests during the construction of the industrial concrete floor (Figure 10.11). The first step in this stage is to control the consistency of the concrete mix required in the design. Next, the required thickness of the flexible material used to form the perimeter expansion joints should be checked by a direct visual method. Then, while laying the concrete mix, the thickness of the laid concrete layer and the evenness of the top surface of the concrete industrial floor should be successively checked (controlled) using a non-destructive geodetic method. The unevenness of the upper surface is acceptable within the permissible upper and lower deviation, the values of which are given in Figure 10.1. However, the level of the top surface must be as designed. The next step is to control the incised antispasmodic expansion slits in terms

FIGURE 10.12   View: (a) of the process of cutting the expansion joint, (b) testing the depth of the cut anti-contraction expansion joint using the direct visual method.

of: their completion date, maintaining the appropriate size of the dilated field, the length of the dilated field, and the aspect ratio of the dilated field. The depth and width of these fissures should then be tested using the direct visual method (Figure 10.12). After performing these tests and confirming the correctness of the expansion joints, it is possible to fill these joints with elastic material.

When diagnosing an existing industrial concrete floor, which is characterized by filled expansion joints, the depth of these joints can be examined using the non-destructive ultrasound or impact-echo method, which excludes the occurrence of the damage of the filling in many places. In tests carried out with the ultra-sonic method, an ultrasonic probe with cylindrical heads is used for this purpose (Figure 10.13). First, the ultrasonic heads should be set on the floor surface with a spacing of 400 mm in a place with no visible cut expansion gaps and cracks. Afterwards, the time in which the ultrasonic wave in the concrete passes from the transmitting to the receiving head should be measured in order to determine its speed. The heads should then be set at the same spacing, but at a distance of 200 mm from each head to the cut expansion joint, in order to determine the depth of the expansion joint.

Stage 3 includes the tests of the concrete industrial floor (Figure 10.14) before it is put into use, or on the existing floor to be diagnosed. The first diagnostic step in stage 3 is thickness control using non-destrictive ultrasonic tomography (Figure 10.15). Testing with this method should be carried out in the same way as described in Chapter 8.

Then, the evaluation of the near-surface tensile strength of the industrial concrete floor should be carried out using the non-destructive pull-off method in randomly

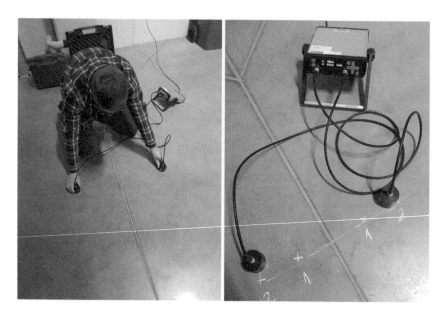

FIGURE 10.13 Examination of the depth of a cut expansion joint using the ultrasonic method in an industrial concrete floor.

made places. The test should begin with the selection of representative measurement sites and preparation of the surface. In each of these places, an incision at least 15 mm deep in the floor should be made, where a steel measuring disc with a diameter of 50 mm is then placed. The disc should then be detached from the floor, and the tensile strength value should be determined next to the surface.

The depth of cracks in the industrial concrete floor, if there are any, should be examined using the non-destructive ultrasonic method in accordance with the methodology provided in stage 2 for testing the depth of cut antispasmodic expansion joints

Then, a minimum of six core samples with a diameter of 150 mm and a height equal to the thickness of the floor should be obtained from the concrete floor in order to test the compressive strength of the concrete along the thickness of the floor. In addition, a small piece of concrete must be cut to obtain at least three specimens 71 mm long, 71 mm wide, and as high as the thickness of the industrial concrete floor in order to test the abrasion in accordance with the standard [10]. In the case of newly constructed floors, all tests should be performed after 28 days.

For ultrasonic testing of core samples, it is proposed to use an ultrasonic probe with special exponential heads with a frequency of 40 kHz and point contact with the tested surface, with measuring points being placed on the side surfaces of the samples with a spacing of 5 mm (Figure 10.16).

After performing ultrasonic tests, the core samples should be cut so that their height is 150 mm, and then subjected to compressive strength tests in a testing machine. Based on the performed ultrasonic tests, a correlation relationship (or possibly a hypothetical relationship from the literature) should be developed for the

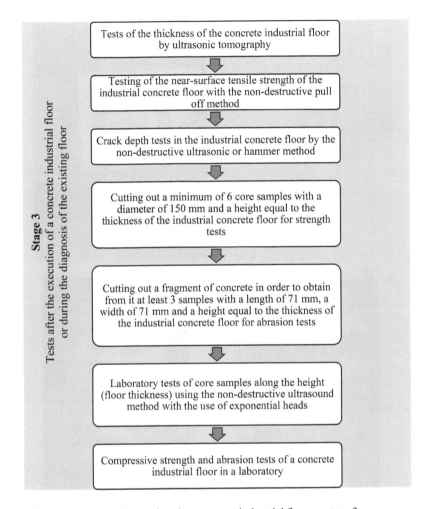

Stage 3

Tests after the execution of a concrete industrial floor or during the diagnosis of the existing floor

Tests of the thickness of the concrete industrial floor by ultrasonic tomography

Testing of the near-surface tensile strength of the industrial concrete floor with the non-destructive pull off method

Crack depth tests in the industrial concrete floor by the non-destructive ultrasonic or hammer method

Cutting out a minimum of 6 core samples with a diameter of 150 mm and a height equal to the thickness of the industrial concrete floor for strength tests

Cutting out a fragment of concrete in order to obtain from it at least 3 samples with a length of 71 mm, a width of 71 mm and a height equal to the thickness of the industrial concrete floor for abrasion tests

Laboratory tests of core samples along the height (floor thickness) using the non-destructive ultrasound method with the use of exponential heads

Compressive strength and abrasion tests of a concrete industrial floor in a laboratory

FIGURE 10.14   Methodology of testing concrete industrial floors – stage 3.

concrete embedded in the industrial floor under test between the longitudinal ultrasonic wave velocity and the compressive strength, as described in Chapter 3. This relationship will be used to identify the value of the compressive strength of the concrete along the thickness of the tested floor (see Figure 10.8). The compressive strength in the upper zone of the industrial concrete floor should not be less than 10% when compared to that in the middle zone. Abrasion tests should be performed in accordance with the standard [10]. The abrasion test involves measuring the change in height and weight of a cubic sample with a side of 71 mm. The sample is placed on a disc sprinkled with abrasive powder, and is then pressed and put into rotation with the help of a grinding machine [10].

Finally, it should be added that the compressive strength, near-surface tensile strength, and abrasion resistance of the industrial concrete floor should not be less than those required by the floor designer.

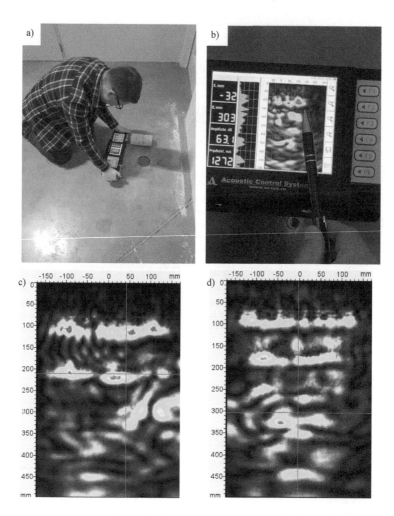

FIGURE 10.15 Exemplary tests of the thickness of the industrial concrete floor using the non-destructive ultrasonic tomography method: (a) conducting research, (b) current analysis of the obtained results (imaging), (c) exemplary imaging of fragments of the industrial concrete floor with a thickness of about 200 mm, (d) exemplary imaging of fragments of the floor with a thickness of about 160 mm.

10.4 CONCLUSIONS

The defectiveness of industrial floors commonly used in industrial and warehouse facilities in many cases results in their exclusion from use in order to perform the necessary repairs. This chapter lists and discusses common disadvantages of these floors, illustrating them with examples from civil engineering practice and giving reasons for their occurrence. It was pointed out that the lack of effective control of the execution of floors had a large share in the formation of these defects. In order to counteract this shortcoming, the authors, based on their own experience acquired in

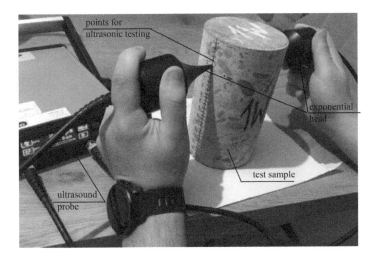

FIGURE 10.16   View of the core sample tested using the ultrasonic method with the use of exponential heads.

construction practice, have developed and included in this chapter an original methodology that is also useful in non-destructive testing of newly built floors before they are put into use and in diagnosing already used floors.

REFERENCES

1. Mynarčík, P. (2013). Technology and trends of concrete industrial floors. Procedia engineering, 65, 107–112.
2. Hedebratt, J. (2012). Industrial fibre concrete floors. TRITA-BKN. Bulletin, 113, 90.
3. Marushchak, U.D., Sydor, N.I., Braichenko, S.P., Margal, I.V., & Soltysik, R.A. (2019, December). Modified fiber reinforced concrete for industrial floors. IOP conference series: Materials science and engineering, 708(1), 012094.
4. Neal, F.R. (2002). Concrete industrial ground floors. London, UK: Thomas Telford Publishing.
5. Vitt, G. (2003). Steel fibre concrete industrial floors. In International RILEM workshop on test and design methods for steelfibre reinforced concrete (pp. 201–208). RILEM Publications SARL.
6. Tatnall, P.C., & Kuitenbouwer, L. (1992). Steel fiber reinforced concrete in industrial floors. Concrete international, 14(12), 43–47.
7. Silfwerbrand, J.L., & Farhang, A.A. (2014). Reducing crack risk in industrial concrete floors. ACI materials journal, 111(6), 681.
8. Medeliene, V., & Žiogas, V.A. (2010). Making solutions for choosing industrial concrete floors and expedience of reliability evaluation. Journal of civil engineering and management, 16(3), 320–331.
9. Stawiski, B., & Radzik, Ł. (2017, October). Need to identify parameters of concrete in the weakest zone of the industrial floor. IOP conference series: Materials science and engineering, 245, 2, p. 022063.
10. EN 13892-3 Methods of test for screed materials – Part 3: Determination of wear resistance - Böhme.

11 The Diagnostics of Polyurethane-Cement Floors

11.1 INTRODUCTION

Polyurethane-cement composites are more and more commonly used to make the top layer of concrete industrial floors [1–5]. This is the case in large-scale facilities where it is required to ensure a higher standard in terms of hygienic conditions [6, 7]. An example of such facilities are warehouses for medical and food products, etc. As a rule, these composites are characterized by a material composition that is prepared and controlled in factory conditions [8]. In this way, the consistency of the composition is ensured, and possible errors in the dosing of the composite components are limited to a minimum. Afterwards, under construction conditions, the process of laying the resulting mixture is carried out after the previous mixing of the dry ingredients with polyurethane resin. This process should comply with the conditions and requirements specified by the manufacturer, because then the floor is guaranteed to meet the technical requirements for such floors [9–13]. The most important of these requirements are given in Figure 11.1.

Bearing in mind Chapter 10, which is devoted to the diagnostics of industrial concrete floors, an attempt was made in this chapter to not duplicate what was written there. In relation to top layers made of polyurethane-cement composites, a frequent disadvantage is the lack of compliance of the thickness with the designed thickness. As a rule, this thickness is too small, and often considerably, which results in wearing out in a shorter time than that assumed. Another important disadvantage is the susceptibility to the formation on the surface of the top layer of local abrasions and dark discolorations, local roughening, and even deep damage reaching up to the concrete substrate. The amount of this damage increases with the period of operation. The consequence of this significant disadvantage is not only the deterioration or loss of the "hygienic" standard, but most of all the significantly reduced durability of the top layer made of polyurethane-cement composite.

11.2 SIGNIFICANT DEFECTS AND REASONS FOR THEIR FORMATION

In line with what was said above, the disadvantage of many concrete industrial floors with a polyurethane-cement top layer is that the thickness of the top layer is too small in relation to the designed one. Such problems are found during the diagnosis of these floors. An example can be seen in one of the large-area floors investigated by the authors (Figure 11.2), from which core samples were taken for testing. Based

DOI: 10.1201/9781003288374-11

Thickness of the polyurethane-cement top layer	• According to the design, most often from 4.5 to 6 mm
The conditions for starting the floor	• After installing window and door joinery in the building, • In the absence of drafts drying the laid floor, • Without the possibility of intense sunlight or heating of the laid floor, • At an air temperature in the building from + 10°C to + 40°C and relative humidity up to 65%, • At a substrate temperature of + 10°C to + 40°C and bulk moisture up to 4%.
Requirements for the concrete substrate	• Compressive strength of a minimum of 25 MPa, • A minimum pull-off strength of 1.5 MPa, • The surface should be clean, strong and dry, or moistened to a mat-damp condition, • The surface should be cleaned of unbound particles, such as: oil, grease, old coatings, surface care products, • The surface evenness should be preserved.
Requirements for priming the concrete substrate	• Priming the concrete substrate with a two-component epoxy primer resin for priming with full broadcast of quartz sand with granulation 0.4–0.7 mm
Mechanical properties of the polyurethane-cement top layer	• Compressive strength according to the design (usually at least 50 MPa*), • Bending strength according to the design (usually at least 15 MPa*) • Tensile strength according to the design (usually at least 9 MPa*), • Pull-off strength according to the project (usually a minimum of 2.5 MPa; leading to failure in the concrete substrate; the value of a single test - a minimum of 1.5 MPa), • Abrasion resistance according to the design (usually a maximum of 4.58 cm^3 / 50 cm^2).
Chemical resistance	• Chemical resistance according to the design (most often required is resistance to organic and inorganic acids, alkalis, amines, salts and solvents)
Thermal shock resistance	• Thermal shock resistance according to the design (most often resistance to thermal shock up to + 70 °C at 6 mm thickness - no cracks and / or delamination)
Temperature of service	• At an air temperature from -40°C to + 90°C and relative air humidity of of a maximum of 85%
Cracks, delamination	• They are not allowed

*values achieved after 28 days at an ambient temperature of + 23 °C and a relative air humidity of 50%

FIGURE 11.1 Summary of the most important requirements for concrete industrial floors with a polyurethane-cement top layer.

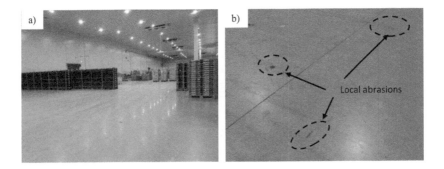

FIGURE 11.2 View: (a) general view of the diagnosed floor, (b) numerous local floor abrasions [14].

on measurements using the geometric method, it was found that the thickness of the polyurethane-cement top layer in this floor is in the range from 3.03 mm to 4.46 mm (average 3.83 mm), which is a shortage of 23.4% in relation to the thickness design of 5 mm. Such a large shortage of thickness prevents the floor from being approved for use. The results are illustrated by the diagram presented in Figure 11.3. The reason for this defect is the contractor's unreliability and the lack of effective control by the technical supervision at the execution stage.

Strength parameters of concrete industrial floors with a polyurethane-cement top layer, such as the compressive strength of the concrete substrate and, above all, the pull-off strength of the top layer and its abrasion, are not tolerated to be lower than those required. Figure 11.4 shows an example of the damaged surface of the top layer due to reduced abrasion resistance, and also an example of a local loss of the

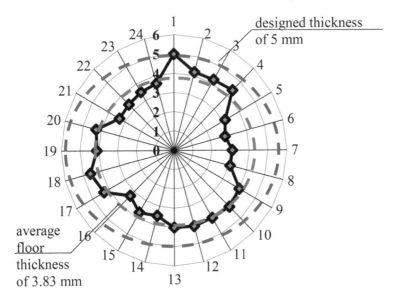

FIGURE 11.3 Results of measurements of the thickness of the polyurethane-cement top layer in the tested industrial concrete floor.

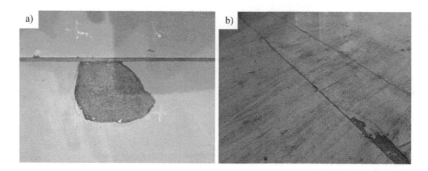

FIGURE 11.4 Examples: (a) local loss of the polyurethane-cement top layer, which is characteristic for places with a reduced pull-off strength to the concrete substrate, (b) a worn top polyurethane-cement layer due to its reduced abrasion.

polyurethane-cement top layer. Such a loss is characteristic for places with a reduced resistance to detachment from the concrete substrate. The reason for the reduced pull-off strength, either locally or over the entire surface, is usually the failure of the concrete substrate and the primer requirements, including, in particular, the lack of full quartz sand sprinkles (detailed in Figure 11.1). Civil engineering practice also shows that a possible cause is a lack of primer on the concrete substrate. Effective technical supervision can eliminate these causes. On the other hand, the causes of reduced abrasion resistance may be caused by many things, for example, incorrect dosing of the polyurethane resin with the dry ingredients, and the failure to meet the required conditions (given in Figure 11.1) for starting the production of the floor.

As shown in Figure 11.1, industrial concrete floors with a polyurethane-cement top layer are characterized by resistance to thermal shock of up to +70°C. This should be understood as a lack of such resistance at higher temperatures. However, in practice, and also during use, in many buildings the temperature of +70°C in the top layer is exceeded. The lack of resistance to temperatures above +70°C is manifested by the formation of both abrasions and dark discoloration on the surface, and also local roughness and burns of the surface of the top layer, as illustrated in Figure 11.5. This happens when the wheels of driving forklifts move on the floor.

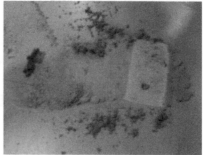

FIGURE 11.5 View of damage in the form of abrasions and dark discoloration as a result of overburning of the polyurethane-cement layer.

TABLE 11.1

Sample Results of the Pull-off Strength of the Polyurethane-Cement Top Layer from the Concrete Substrate after the Forklift Drive Wheel Spinning Test

The Value of the Pull-off Strength f_b Directly at the Place of the Spinning Test of the Forklift Drive Wheel	The Value of the Pull-off Strength f_b Next to the Place Where the Forklift Drive Wheel Spinning Test was Performed	Percentage Decrease in the Value of the Pull-off Strength f_b at the Point of Spinning of the Forklift Drive Wheel
1.22 MPa	1.68 MPa	28.17 %

This has been proven on the basis of the authors' research (conducted during attempts to spin the wheels of these forklifts) described in [14], which undoubtedly shows that after just 3 seconds of the forklift drive wheel spinning, regardless of whether it is loaded or not, the surface temperature of concrete industrial floors with a polyurethane-cement top layer is very quickly close to 70°C. After 9 seconds it is almost 100°C, and after 12 seconds it is over 120°C. There is an unequivocal conclusion from this that floor with a polyurethane-cement top layer should be used in such a way that there is no spinning of the wheels of the means of transport that travel over them. This should be strictly specified in the user manual.

The consequence of the damage caused by the lack of resistance to thermal shock exceeding the temperature of 70°C on the floor surface is the reduction of the pull-off strength of the polyurethane-cement top layer to the concrete substrate in the damaged places, which was demonstrated on the basis of tests, the results of which are presented in Table 11.1.

As shown in Table 11.1, the consequence of local abrasions and burns on the top polyurethane-cement layer is a significant loss of its pull-off strength, reaching almost 30%, compared to the pull-off strength determined in the undamaged parts of the top layer. This results in wear of the top layer similar to that shown in Figure 11.1.

11.3 METHODOLOGY OF TESTING CONCRETE INDUSTRIAL FLOORS WITH A POLYURETHANE-CEMENT TOP LAYER

Following the example of previous chapters in this chapter, an original methodology for testing industrial concrete floors with a polyurethane-cement top layer has also been developed. It is shown graphically in Figures 11.6–11.8, and is described below. Figure 11.6 shows a general diagram, and Figures 11.7 and 11.8 show a detailed diagram of the methodology. This methodology may be useful not only for newly constructed floors, but also for floors diagnosed during operation.

Stage 1 includes tests prior to commencing the production of an industrial concrete floor with a polyurethane-cement top layer (Figure 11.7). It is necessary to check whether the building has built-in window and door joinery. Then, the temperature and relative air humidity in the building should be measured, which, according to the requirements given in Figure 11.1, should be in the range from +10°C to +45°C

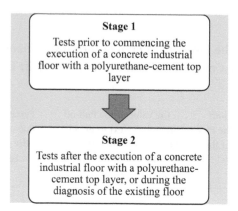

FIGURE 11.6 General diagram illustrating the methodology of testing industrial concrete floors with a polyurethane-cement top layer.

for temperature and up to 65% for humidity. The next step is to test the temperature and bulk moisture content of the concrete substrate on which the polyurethane-cement top layer will be laid. According to Figure 11.1, it is usually required that the substrate temperature is in the range of +10°C to +40°C, and the bulk moisture content is not higher than 4%. Then, using a non-destructive geodetic method, the level and evenness of the upper surface of the concrete substrate should be checked. Unevenness of this surface is acceptable, but only with the admissible upper and lower deviation. Typically, they total a maximum of 6 mm. If these conditions are met, the polyurethane-cement top layer can be laid.

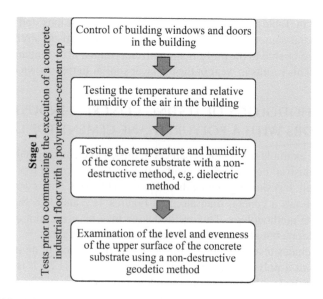

FIGURE 11.7 Methodology of testing concrete industrial floors with a polyurethane-cement top layer – stage 1.

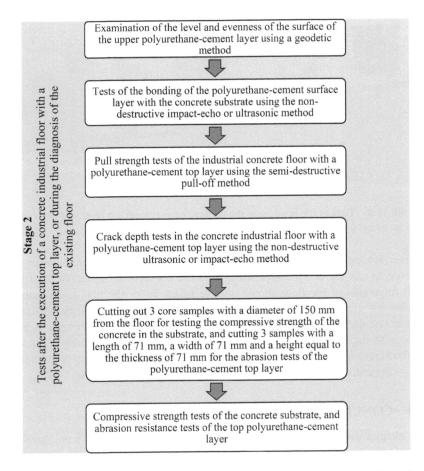

FIGURE 11.8 Methodology of testing concrete industrial floors with a polyurethane-cement top layer – stage 2.

The first diagnostic step in stage 2 (Figure 11.8) is to test the evenness of the floor using the geodetic method, and then to conduct the bond testing of the polyurethane-cement top layer with the concrete substrate using the impact-echo or ultrasonic method (Figure 11.9). In newly constructed floors, tests of the bonding of the top layer with the substrate should be performed 7 days after the top layer has been made.

The next step is to determine the pull-off strength of a concrete industrial floor with a polyurethane-cement top layer using the semi-destructive pull-off method in randomly made places. In the case of newly constructed floors, these tests should be carried out after 28 days, as described in Chapter 12 in Stage 3. The depth of cracks in the floor, if there are any, should then be examined using the non-destructive ultrasonic or impact-echo method in accordance with the methodology given in Chapter 10 in Stage 2 concerning the examination of the depth of cut antispasmodic expansion joints.

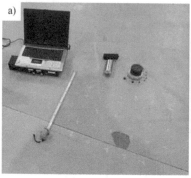
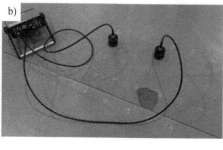

FIGURE 11.9 Tests of the bonding of the polyurethane-cement top layer with the concrete substrate using: (a) impact-echo method, (b) ultrasonic method.

Then, at least three core samples with a diameter of 150 mm should be taken from the floor to test the compressive strength of the concrete substrate. At least three small fragments of the floor should also be taken. From the obtained fragments, samples that are 71 mm long, 71 mm wide, and 71 mm high should be obtained in order to test the abrasion resistance of the top polyurethane-cement layer in accordance with the standard [15]. The compressive strength of the concrete substrate and the abrasion resistance of the industrial concrete floor with a top polyurethane-cement layer should not be lower than those specified by the designer.

11.4 CONCLUSIONS

This chapter discusses the disadvantages of concrete floors with a polyurethane-cement top layer. They are illustrated with examples and exemplary research results taken from my own construction practice. The reasons for the occurrence of these defects were indicated. Much space has been devoted to one of the significant drawbacks resulting from improper use of these floors. This chapter also includes the methodology of testing these floors with the use of non-destructive methods, useful for newly constructed floors as well as those diagnosed during operation.

REFERENCES

1. Ryszkowska, J., Rębiś, J., & Langner, M. (2007). Polyurethane coating for protecting concrete floors using the system with fillers produced by recycling. Archives of civil and mechanical engineering, 7(1), 53–60.
2. Carneiro, A.F.B., Daschevi, P.A., Langaro, E.A., Pieralisi, R., & Medeiros, M.H.F. (2021). Effectiveness of surface coatings in concrete: Chloride penetration and carbonation. Journal of building pathology and rehabilitation, 6(1), 1–8.
3. Hussain, H.K., Liu, G.W., & Yong, Y.W. (2014). Experimental study to investigate mechanical properties of new material polyurethane–cement composite (PUC). Construction and building materials, 50, 200–208.
4. Das, A., & Mahanwar, P. (2020). A brief discussion on advances in polyurethane applications. Advanced industrial and engineering polymer research, 3(3), 93–101.

5. Mundo, R.D., Labianca, C., Carbone, G., & Notarnicola, M. (2020). Recent advances in hydrophobic and icephobic surface treatments of concrete. Coatings, 10(5), 449.
6. Carpentier, B. (2005). Improving the design of floors. In Handbook of hygiene control in the food industry (pp. 168–184). Cambridge, UK: Woodhead Publishing.
7. Yilin, P., Wenhua, Z., Wanting, Z., & Yunsheng, Z. (2022) Study on the mechanical behaviors and failure mechanism of polyurethane cement composites under uniaxial compression and tension. Archives of civil and mechanical engineering, 22, 18. https://doi.org/10.1007/s43452-021-00342-z
8. Somarathna, H.M.C.C., Raman, S.N., Mohotti, D., Mutalib, A.A., & Badri, K.H. (2018). The use of polyurethane for structural and infrastructural engineering applications: A state-of-the-art review. Construction and building materials, 190, 995–1014
9. Maj, M., Ubysz, A., & Tamrazyan, A. (2018). Durability of polyurethane-cement floors. MATEC Web of Conferences (Vol. 251, p. 02026). EDP Sciences.
10. Vipulanandan, C., & Liu, J. (2005). Performance of polyurethane-coated concrete in sewer environment. Cement and concrete research, 35(9), 1754–1763.
11. Somarathna, H.M.C.C., Raman, S.N., Mohotti, D., Mutalib, A.A., & Badri, K.H. (2021). Behaviour of concrete specimens retrofitted with bio-based polyurethane coatings under dynamic loads. Construction and building materials, 270, 121860.
12. Pan, X., Shi, Z., Shi, C., Ling, T.C., & Li, N. (2017). A review on surface treatment for concrete–Part 2: Performance. Construction and building materials, 133, 81–90.
13. Almusallam, A.A., Khan, F.M., Dulaijan, S.U., & Al-Amoudi, O.S.B. (2003). Effectiveness of surface coatings in improving concrete durability. Cement and concrete composites, 25(4–5), 473–481.
14. Sadowski, Ł., Hoła, J., Żak, A., & Chowaniec, A. (2020). Microstructural and mechanical assessment of the causes of failure of floors made of polyurethane-cement composites. Composite structures, 238, 112002.
15. EN 13892-3 Methods of test for screed materials – Part 3: Determination of wear resistance – Böhme.

12 The Diagnostics of Cement Screeds

12.1 INTRODUCTION

One of the conditions for the flawless execution of cement screeds is compliance with the technological and technical requirements (established by the designer) contained in the applicable standards [1–6]. The most important of the requirements are summarized in Figure 12.1.

However, it is also necessary to control the entire production process on an ongoing basis by carrying out technical supervision with the use of non-destructive and semi-destructive methods [7–10]. For the control of the screeding process to be effective, it should be methodical. A lack of such control can result in defective cement screeds. Construction practice shows that the frequent, significant defects of cement screeds include, first of all, a thickness that is inconsistent with that which is designed, too low strength parameters in relation to those required, the unacceptable occurrence of deep and surface cracks, shallowly cut expansion joints, and no circumferential expansion from load-bearing walls.

12.2 EXAMPLES OF DEFECTS

A significant disadvantage that is often found when diagnosing cement screeds is a thickness that is not in accordance with that which is designed. Let one of the cement screeds investigated by the authors serve as an example (Figure 12.2). Using the results shown in Figure 12.3, it was found that the thickness for this screed, which should be 60 mm, differs in individual rooms of the building and falls within a wide range of between 28 and 70 mm, meaning that the maximum thickness shortage is over 50% [11].

The strength parameters of cement screeds (such as compressive strength, flexural strength, and tensile strength) that are lower than those required are not tolerated. This is often the reason for removing a screed that does not meet these requirements. In such circumstances, the screed needs to be made it again. Exemplary values of the mechanical parameters obtained in work [11] for one of the cement screeds shown in Figure 12.2, in comparison with the requirements given in Figure 12.1, are summarized in Table 12.1.

As was stated in Figure 12.1, the values of all tested strength parameters are lower than required. On the other hand, the value of the near-surface tensile strength is definitely too low. This is due to the too low compressive strength of the upper zone of the screed, which was found on the basis of research. Figure 12.4 shows the course of the compressive strength of the cement mortar along the thickness of the cement screed, which was obtained using the non-destructive ultrasonic

DOI: 10.1201/9781003288374-12

Screed thickness	• According to the design, most often 40–60 mm
Conditions for the commencement of the execution of the screed	• After installing windows and doors in the building, • In the absence of drafts drying the mortar, • No possibility of intense sunlight or heating of the applied mortar, • At an air temperature from + 10 °C to + 25 °C and a relative humidity from 65% to 95%, • At a substrate temperature from + 10° C to + 25 °C and a bulk moisture up to 4%.
Preparation of the cement mix	• Mechanical mixing - without homogenising the mixture, or - with homogenization of the mixture according to provisions in the requirements of the mixture manufacturer.
Unevenness of the upper surface of the concrete substrate or ceiling	• It is required not to exceed the deviations noted in [1–6]
Execution of peripheral expansion joints	• For the full thickness of the floor, made of non-absorbent flexible foam, • The width of the slots is at least 7 mm.
Execution of anti-shrinkage incised expansion joints	• A time limit for making the joints - in the first 24 hours after the floor has been made, • The maximum size of a dilated field - 36 m², • The maximum length of the dilated field - 6 m, • The maximum proportions of the sides of the dilated field - 1: 1.5, • The minimum depth of cut expansion joints from 1/3 to 1/2 of the floor thickness, at right angles, • Gap widths from 3 to 5 mm.
Mechanical properties	• Surface tensile strength according to the design (usually at least 1.5 MPa), • Compressive strength according to the design (usually at least 20 MPa), • Flexural strength according to the design (usually at least 5 MPa), • Abrasion resistance according to the design (usually a maximum of 22 cm³ / 50 cm²).
Cracks	• They are not allowed
Unevenness of the top surface of the floor	• It is required not to exceed the deviations recorded in [1–6]

FIGURE 12.1 A list of the most important technological and technical requirements for the implementation of cement screeds.

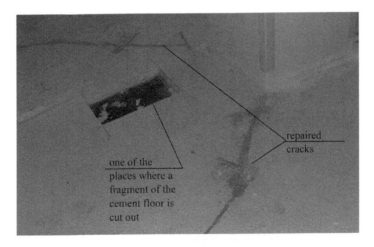

FIGURE 12.2 View of the screed to be assessed with an example of the cutout of a fragment of the screed to be tested.

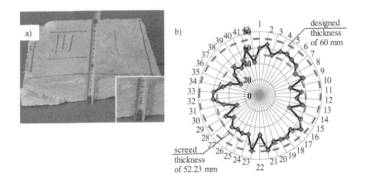

FIGURE 12.3 Examples of the results of testing the thickness of one of the cement screeds: (a) fragment cut out for testing, (b) test results.

TABLE 12.1

Comparison of the Mechanical Parameters for the Tested Cement Screed with the Required Values

The Tested Mechanical Parameter of the Cement Screed	The Value of the Mechanical Parameter Obtained in [11]	The Required Value of the Mechanical Parameter
Near-surface tensile strength	0.55 MPa	1.5 MPa
Compressive strength	15.1 MPa	20.0 MPa
Flexural strength	2.3 MPa	5.0 MPa

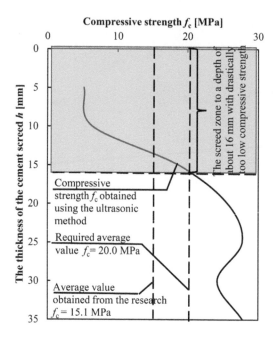

FIGURE 12.4 The course of the compressive strength of the cement mortar along the thickness of the cement screed, determined using the non-destructive ultrasonic method.

method. In Figure 12.4, the required average compressive strength of 20 MPa and the average strength obtained from tests of 15.1 MPa are marked with dashed vertical lines, while the continuous non-vertical line shows the values obtained using the ultrasonic method.

It can be seen from Figure 12.4 that up to a depth of about 16 mm, counting from the top surface of the screed, the compressive strength is lower than the required strength, and in the near-surface zone with a thickness of about 10 mm, it is far too low.

Cracks are also a significant disadvantage of cement screeds. They are not acceptable, and are found very often, as illustrated in Figure 12.5. They can have a length of up to several meters and an opening of up to 1 mm. They can go through, or nearly through, the entire thickness. Their course is most often multidirectional or parallel to the lines of the incorrectly cut antispasmodic expansion joints in the screed. This anomaly is due to making the slots too late, too large expansion slots, too long expansion slots, exceeding the recommended maximum proportions of the sides of the expansion slots, and a too small depth of the cut expansion slots. Examples of these irregularities are illustrated in Figure 12.6. This is the result of non-compliance with the technological and technical requirements listed in Figure 12.1, as well as the lack of effective methodological control by the technical supervision.

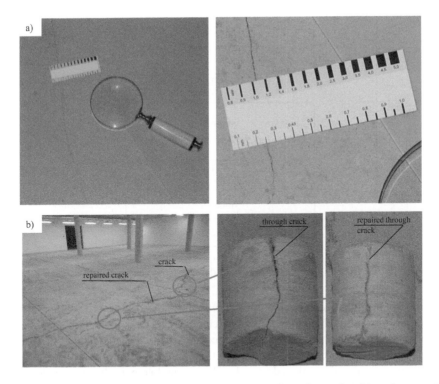

FIGURE 12.5 An exemplary view of cracks on the surface of screeds with a close-up of small cut out core samples: (a) measurement of the crack width using a magnifying glass and a comparative template, (b) proof of the crack's thoroughness.

12.3 METHODOLOGY OF TESTING CEMENT SCREEDS

On the basis of the above-mentioned defects of cement screeds, attention was paid to the fact that they arise as a result of non-compliance with the technological and technical requirements at the stage of making them, as well as ineffective control

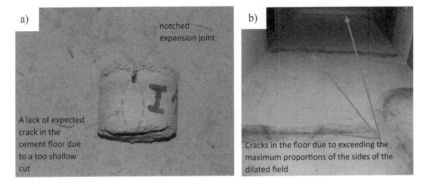

FIGURE 12.6 View: (a) behind a shallow expansion joint, (b) cracks due to exceeding the maximum proportions of the sides of the dilated field.

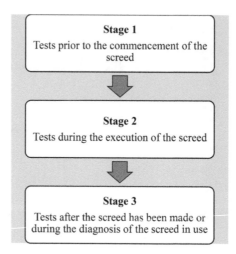

FIGURE 12.7   General diagram illustrating the developed test methodology for cement screeds.

during the technical supervision. This is one of the reasons why the authors developed an original methodology for testing cement screeds. Knowing this methodology and applying it in practice can help to reduce defects. The methodology may also be useful for testing new screeds prior to their commissioning, and may be useful in diagnosing already exploited screeds. This methodology is shown graphically in Figures 12.7 and 12.8, and is described below. Figure 12.7 shows an overview of the methodology, and Figures 12.8 and 12.9 show a detailed diagram of it.

Stage 1, in terms of checking the incorporation of window and door joinery in the building and testing the air temperature and relative humidity, is identical to the

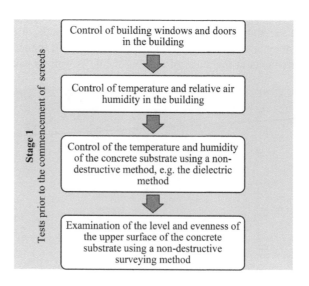

FIGURE 12.8   Methodology of testing cement screeds – stage 1.

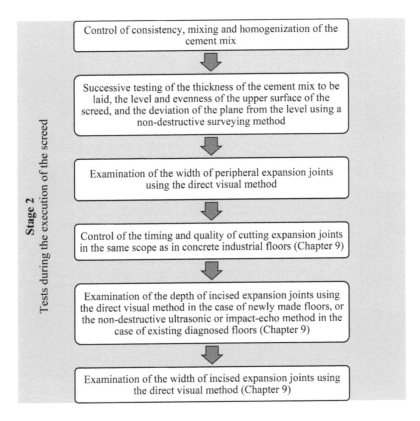

FIGURE 12.9 Methodology of testing cement screeds – stage 2.

test of concrete industrial screeds described in Chapter 9. As cement screeds are primarily laid on a concrete substrate, the last step in this stage is testing the temperature and mass humidity of the substrate on which the screed will be installed. According to Figure 12.1, it is usually required that the temperature of the substrate is in the range of +10°C to +25°C, with the mass moisture content of the substrate not being higher than 4%. The final step in this stage is to test the level and evenness of the top surface of the concrete substrate using a non-destructive surveying method. Unevenness of the upper surface of the substrate is acceptable, but the permissible upper and lower deviation must be maintained. Typically, they amount to a maximum of 6 mm (in total) – according to [1–6].

It is then possible to move on to the second stage, which is followed by carrying out tests during the screeding process (Figure 12.9). These tests do not differ significantly from those performed in stage 2 for industrial concrete screeds, as described in Chapter 10, and should be completed with tests of the depth and width of the cut anti-contraction expansion joints.

Stage 3 includes tests that are carried either after the screed has been made and before it is put into use (Figure 12.10), or during the diagnosis of a screed that is already in use.

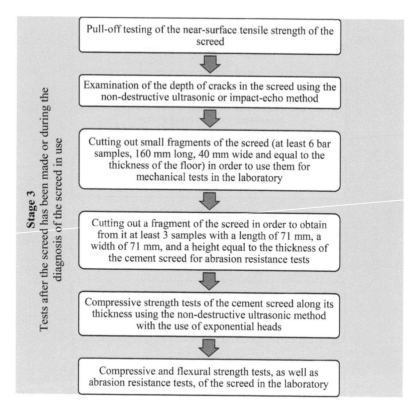

FIGURE 12.10   The methodology of testing cement screeds – stage 3.

The first diagnostic step in stage 3 is the semi-destructive pull-off strength testing of the near-surface tensile strength using the pull-off method. In the case of newly made screeds, these tests should be performed after 28 days. The study should begin with the selection of measurement sites and surface preparation. Then, in these places, incision should be made at least 15 mm deep in the screed and a steel measuring disc with a diameter of 50 mm should be attached. The disc should then be detached from the screed using the pull-off method, and the value of the near-surface tensile strength should be determined. Figure 12.11 shows a view of the near-surface tensile strength tests carried out using the pull-off method, and also a close-up of a fragment of the screed surface after the tests. The depth of the cracks in the screed should then be examined using the non-destructive ultrasonic or impact-echo method in accordance with the methodology provided and described in stage 2.

Afterwards, small fragments of the screed should be cut out. Their size should enable at least six bar samples that are 160 mm long, 40 mm wide, and which are equal to the thickness of the screed to be obtained. These samples are needed in order to test the compressive strength of the cement mortar along the thickness of the screed, as well as to test the flexural and compressive strength. Moreover, a small fragment of the screed should be cut out in order to obtain at least three samples

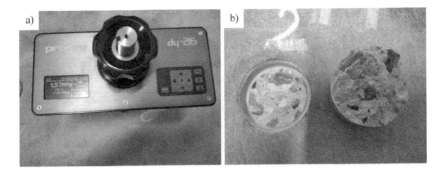

FIGURE 12.11   View: (a) tests of near-surface tensile strength carried out using the pull-off method (b) close-up of a fragment of the screed surface after testing of the near-surface tensile strength.

(71 mm long, 71 mm wide, and as high as the screed thickness) for the purpose of testing abrasion resistance in accordance with the standard [12].

First, the bar samples should be subjected to ultrasonic tests along the direction of the concreting of the screed. For ultrasonic tests, it is proposed to use an ultrasonic probe with special exponential heads with a frequency of 40 kHz and point contact with the tested surface. Measurement points should be placed on the side surfaces of the samples in three rows with a spacing of 5 mm (Figure 12.12).

After performing the ultrasonic tests, the lower zone of the bar samples should be cut so that their thickness is 40 mm. The samples should then be subjected to flexural strength and compression tests in a testing machine [13]. Based on the performed

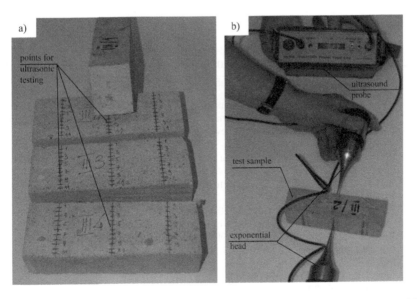

FIGURE 12.12   View: (a) bar samples cut from the cement screed, which were prepared for ultrasonic testing, (b) bar samples tested using the ultrasonic method.

ultrasonic tests, a correlation relationship between the velocity of the longitudinal ultrasonic wave and the compressive strength of the cement mortar should be developed for the tested screeds (it is also possible to choose a hypothetical one from the literature), as was already explained in Chapter 10. This relationship will be used to identify the value of the compressive strength of the cement mortar along the thickness of the screed. The compressive strength in the upper zone of the screed should not be less than 10% when compared to that in the middle zone. On the other hand, the abrasion test, as explained in Chapter 10, consists in measuring the change in height and weight of a cubic sample with a side of 71 mm. The sample is placed on a disc sprinkled with abrasive powder, and then pressed and put into rotation with the help of a grinding machine.

The flexural and compressive strength, as well as the abrasion resistance of the screed, should not be lower than those specified by the screed designer.

12.4   CONCLUSIONS

This chapter lists and discusses the disadvantages of cement screeds. Examples of common flaws are included, along with research results from in-house expert practice. It was indicated that these defects are the result of non-compliance with the technological and technical requirements at the execution stage of the floors, as well as ineffective control of the technical supervision. This chapter also contains and discusses the original three-stage methodology of testing these floors, the use of which in practice can help to reduce the defectiveness. This methodology should also be useful in testing new floors before putting them into service and in diagnosing used floors.

REFERENCES

1. DIN 18202 04.97, "Toleranzen im Hochbau Bauwerke; Bauwerke".
2. Loprencipe, G., & Cantisani, G. (2015). Evaluation methods for improving surface geometry of concrete floors: A case study. Case studies in structural engineering, 4, 14–25.
3. Walsh, K.D., Bashford, H.H., & Mason, B.C. (2001). State of practice of residential floor slab flatness. Journal of performance of constructed facilities, 15(4), 127–134.
4. Face, A. (1987). Floor flatness and levelness: The F-number system. Constr. Specifier, 40(April, 4), 124–132.
5. PN-62/B-10145 Concrete and cement mortar floors. Requirements and technical tests upon receipt (in Polish)
6. ASTM 1155M-96 (2001), "Standard test method for determining FF floor flatness and FL floor levelness numbers".
7. Sadowski, Ł., Hoła, A., & Hoła, J. (2020). Methodology for controlling the technological process of executing floors made of cement-based materials. Materials, 13(4), 948.
8. Bijen, J.M.J.M., & Schlangen, E. (1999). Modelling and testing of floor systems with cementitious screeds and adhesives. In Radical design and concrete practices: Proceedings of the international seminar held at the University of Dundee, (pp. 103–114). Scotland, UK: Thomas Telford Publishing.
9. Carruth, W.D. (2019). Evaluation of commercially available screeds for finishing of rapid-setting concrete.

10. Schlangen, E. (1999). Cementitious screeds and adhesives. In Radical design and concrete practices: Proceedings of the international seminar held at the University of Dundee, Scotland, UK on 7 September 1999 (p. 103). Thomas Telford Publishing.
11. Hoła, J., Sadowski, Ł., & Hoła, A. (2019). The effect of failure to comply with technological and technical requirements on the condition of newly built cement mortar floors. Proceedings of the institution of mechanical engineers, part l: Journal of materials: Design and applications, 233(3), 268–275.
12. EN 13892-3 Methods of test for screed materials – Part 3: Determination of wear resistance – Böhme.
13. EN 13892-2:2004, Methods of test for screed materials – Part 2: Determination of flexural and compressive strength.

13 The Diagnostics of Layered Concrete Floors

13.1 INTRODUCTION

In exploited concrete floors, as a result of various environmental impacts, the near-surface zone is destroyed over time. The thickness of this zone may be even several tens of millimetres. For further safe use of such floors, they are subjected to surface repair, consisting in removing the damaged concrete layer and adding a new layer to it. Over-concreting can be made of concrete with a material composition that is similar to that used to make the floor, but it can also be made of special fine-grained repair concrete. After the new layer is over-concreted, the floor becomes a layered floor. Figure 13.1 shows a diagram explaining the surface repair of a concrete floor with the use of over-concreting.

For a layered floor created in this way, in which the over-concreted layer is the top layer, and the floor with topping up is the bottom layer, the inter-layer bonding is of key importance for durability [1]. Therefore, it needs to be assessed before the floor is put into use. The measure of anastomosis is the pull-off adhesion value f_b. In practice, this adhesion is determined using the semi-destructive pull-off method [2]. According to [3], pull-off adhesion value f_b should be at least 1.5 MPa in each test site.

According to [4], however, it is required that in all areas of the floor there should be one test site for an area of no more than 3 m^2, because then there is sufficient probability of the possibility of omitting the areas of the floor that have lower adhesion than that provided for in [3]. The semi-destructive adhesion test, however, involves local damage to the floor in each test site, and then the necessary repair of these places after completion of the tests. For this reason, in practice, the above-mentioned standard requirement is frequently not followed. This is especially true for large-area concrete floors, where even several hundred local damages may be required. Therefore, in practice, the number of test sites in these floors is drastically reduced. This in turn translates into the possibility of overlooking defective areas in which the interlayer bonding does not meet the condition specified in [3], which, due to the durability of surface repaired floors, is not beneficial.

Therefore, it is advisable and possible to use non-destructive methods for the assessment of interlayer bonding, which the authors have shown in their research. Depending on the needs, such an assessment may be limited to zero-one, involving the locating of delamination between layers. Or it can also be full, in the sense of determining the pull-off adhesion value f_b of the layers. Useful for this purpose are the non-destructive methods described in Chapter 5, namely the 3D laser scanning, impulse response, and impact-echo methods.

DOI: 10.1201/9781003288374-13

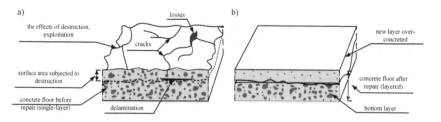

FIGURE 13.1   Diagram explaining the repair of a surface damaged concrete floor: (a) concrete floor before repair, (b) concrete floor after repair.

These methods, used individually, can be used to successfully perform a zero-one assessment of bonding [5]. However, the pull-off adhesion f_b values cannot be determined using these methods if they are used alone. This can be seen in works [6–8], because it is not possible to determine reliable correlation relationships between the pull-off adhesion f_b and the parameters determined separately using individual non-destructive methods.

As can be seen, inter alia, from the authors' own research, with the comprehensive combined use of the above-mentioned non-destructive methods and the use of artificial neural networks (ANNs) with a properly selected structure and a learning algorithm as a tool for associating non-destructive test results, it is possible to assess the pull-off adhesion f_b value of concrete layers in floors. These tests formed the basis for the development of a non-destructive method of assessing this adhesion in surface-repaired concrete floors, which is presented below.

13.2   A NON-DESTRUCTIVE NEURAL METHOD OF ASSESSING PULL-OFF ADHESION

The non-destructive neural method of assessing the pull-off adhesion value of concrete layers in floors was developed in PhD thesis [9, 10] and then published, among others, in [11–21].

It is based on the comprehensive application of three non-destructive methods and ANNs, namely: the 3D laser scanning method, which is used to determine the values of parameters describing the morphology of the bottom surface layer before its over-concreting; the acoustic impulse and impact-echo response methods, which determine the values of parameters on the top surface of the over-concreted layer; and ANNs, which are used as tools for associating parameters obtained with the above-mentioned methods. The 3D laser scanning method and the acoustic impulse response and impact-echo methods are described synthetically in Chapter 5.

For the purpose of developing such an assessment method, a database was built by making and testing model layered concrete floors, for which different (but used in construction practice) methods of treating the surface of the bottom layer were adopted. Therefore, on the surface of the bottom layer, the values of the Sa and Sq parameters, which describe the bottom layer's morphology in several hundred research sites, were determined by means of 3D laser scanning. Then, on the surface of the concrete top layer, the acoustic parameters K_d and N_{av} were determined using

TABLE 13.1

An Exemplary Fragment of a Database Built in [9] Used to Train, Test, and Validate the Artificial Neural Network

	Method Name, Symbol, and Value of Measured Parameters					
	3D Laser Scanning		Impulse Response		Impact-echo	Pull-off
	Sa	Sq	K_d	N_{av}	f_T	f_b
Research Site Number	mm	mm	-	m/s·N	kHz	MPa
1	0.307	0.403	0.058	72.839	8.500	0.870
2	0.317	0.424	0.037	57.736	5.500	0.640
3	0.077	0.241	0.075	43.906	9.500	0.990
4	0.057	0.223	0.071	55.123	9.500	1.020
5	0.359	0.478	0.056	56.002	8.000	0.740
6	0.442	0.582	0.010	89.706	7.000	0.610
7	0.525	0.685	0.009	85.122	8.000	0.780
8	0.592	0.793	0.008	70.897	7.000	0.660
9	0.283	0.369	0.110	39.779	8.500	1.120
10	0.295	0.377	0.072	55.349	8.500	1.040
...	...	...	...	...	...	...
460	0.059	0.401	57.808	0.369	8.500	0.870

the impulse response method, with the f_T parameter being determined using the impact-echo method. These parameters were also determined in several hundred identically distributed research sites. Then, in these places, semi-destructive pull-off tests were performed in order to obtain the pull-off adhesion f_b value needed to train, test and validate the ANN. The database built on the basis of these studies had several hundred sets of parameters, which for selected research points is presented in Table 13.1. The full database is included in [9]. The names of its parameters, in accordance with the nomenclature adopted in Chapter 5 of this book, are as follows:

- Sa – arithmetic mean height,
- Sq – root mean square height,
- K_d – dynamic stiffness,
- N_{av} – average mobility,
- f_T – frequency of reflection of the ultrasonic wave from the bottom of the concrete floor.

On the basis of the analyses carried out in [9], it has been shown that a back propagation ANN with the Quasi-Newton learning algorithm, with the number of neurons of the hidden layer equal to 10, is useful for the non-destructive neural assessment of the bonding of layers in floors. The structure of the network is shown in Figure 13.2.

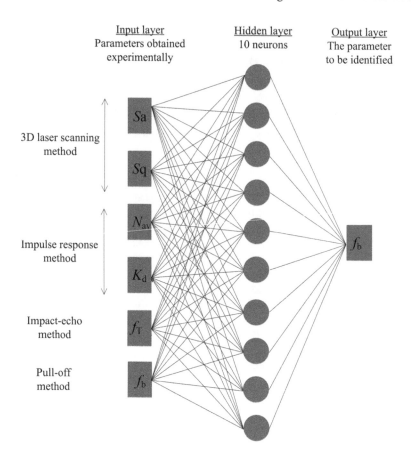

FIGURE 13.2 Structure of a back propagation artificial neural network with the Quasi-Newton learning algorithm, with the number of neurons of the hidden layer equal to 10, which is useful for the non-destructive evaluation of the pull-off adhesion value f_b of the concrete layers in floors.

In the process of training this network, the value of the linear correlation coefficient R equal to 0.9775 was obtained, with 0.9725 being obtained in the testing process, and 0.9481 in the validation process. It allows for the statement that it is possible to provide a reliable neural evaluation of the pull-off adhesion f_b value of the top and bottom concrete layers. This evaluation is based on parameters that are assessed with the following three non-destructive methods: 3D laser scanning, impulse response, and impact-echo methods.

The use of the built database, which is included in [9], as well as the use of the learned, tested and validated (in [9]) back propagation ANN with the Quasi-Newton learning algorithm and the number of neurons of the hidden layer equal to 10 (with the structure given in Figure 13.2) is possible in practice. This was confirmed by experimental verification carried out in laboratory conditions and the practical application documented in [19]. It is possible in situations where the tested layered floor is similar to the one that was used to create the database in [9]. Most important is the

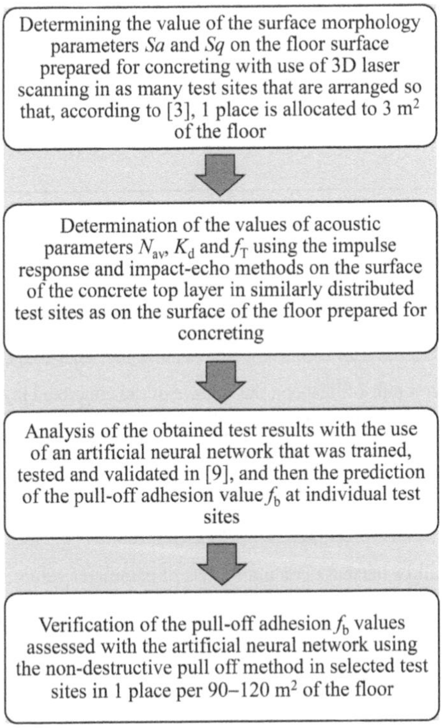

Determining the value of the surface morphology parameters *Sa* and *Sq* on the floor surface prepared for concreting with use of 3D laser scanning in as many test sites that are arranged so that, according to [3], 1 place is allocated to 3 m² of the floor

Determination of the values of acoustic parameters N_{av}, K_d and f_T using the impulse response and impact-echo methods on the surface of the concrete top layer in similarly distributed test sites as on the surface of the floor prepared for concreting

Analysis of the obtained test results with the use of an artificial neural network that was trained, tested and validated in [9], and then the prediction of the pull-off adhesion value f_b at individual test sites

Verification of the pull-off adhesion f_b values assessed with the artificial neural network using the non-destructive pull off method in selected test sites in 1 place per 90–120 m² of the floor

FIGURE 13.3 Methodology of assessing (using non-destructive methods) the values of the pull-off adhesion between the top layer and the bottom layer in the diagnosed layered concrete floors. The methodology uses the database included in [9], and also the trained, tested, and validated ANN.

similarity of the materials of the bottom layer and the over-concreted top layer, as well as the similar thickness of the new over-concreted layer to the one analysed in [9].

The methodology helpful for this application is given in Figure 13.3. In addition, Figure 13.4 shows the developed methodology for creating the database for a different material configuration of the bottom layer and the concrete top layer than the one included in [9].

13.3 METHODOLOGY OF THE NON-DESTRUCTIVE NEURAL EVALUATION OF THE PULL-OFF ADHESION VALUE

In the case of using the database and the trained, tested and validated ANNs in construction practice, as in [9], it is necessary to obtain two parameters (*Sa* and *Sq*) that describe the morphology of the bottom layer surface. These parameters are assessed using 3D laser scanning at each test site before the over-concreting of the bottom layer. After that it is also necessary to obtain three acoustic parameters (K_d, N_{av}, and f_T), which are assessed on the surface of the over-concreted top layer using two non-destructive acoustic methods (the impulse response and impact-echo methods), as shown in Figure 13.3.

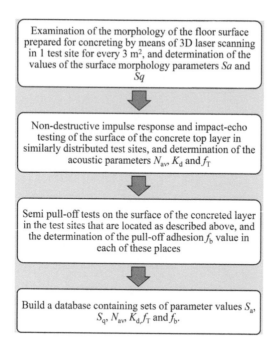

FIGURE 13.4 Methodology of building a database for the purpose of the non-destructive neural assessment of the pull-off adhesion between the top layer and the bottom layer in floors (in the case of a different material configuration of these layers than in [9]).

In the methodology presented in Figure 13.3, one should start with determining the values of the surface morphology parameters Sa and Sq on the floor surface prepared for over-concreting. The values are determined, according to [3], with the use of 3D laser scanning in as many test locations in order for 1 place to be allocated per 3 m^2 of the floor. The next step involves the determination of the values of the acoustic parameters N_{av}, K_d, and f_T using the impulse response and impact-echo methods on the surface of the top layer in test sites that correspond to those on the surface of the floor prepared for concreting. The obtained results should then be analysed with the use of a back propagation ANN with the Quasi-Newton learning algorithm, with the number of neurons of the hidden layer equal to 10. The pull-off adhesion value at individual test sites can then be predicted. The next step is to verify the accuracy of the pull-off adhesion f_b values assessed by the ANN using the non-destructive pull-off method in selected test locations, for example, in 1 place per 90–120 m^2 of the floor.

In turn, the methodology that explains the method of building a database for a material configuration of the bottom layer (subjected to surface repair) and the newly topped layer (e.g., with the use of repair mortars), which is different than the configuration in [9], is given in Figure 13.4.

In the case of the methodology presented in Figure 13.4, it is necessary to start with the implementation of experimental tests of the floor that needs to be surface repaired by over-concreting a new layer. The results of the tests will be used to create a database. First, it is necessary to perform tests using the 3D laser scanning method on the surface of the bottom layer (intended for over-concreting) in 1 test site for every 3 m^2 of the floor. Afterwards, the values of the surface morphology parameters Sa and Sq need to be determined. The next step is to perform impulse response and impact-echo non-destructive acoustic tests on the surface of the newly topped concrete top layer in similarly distributed research sites, with the values of the acoustic parameters N_{av}, K_d, and f_T being determined in these places. Then, pull-off tests should be performed at the same test points, and the pull-off adhesion f_b value should be determined. This research completes the process of building a database with sets of the following parameter values: S_a, S_q, N_{av}, K_d, and f_T.

This database should be used to train and test a back propagation ANN with the Quasi-Newton learning algorithm, with the number of neurons of the hidden layer equal to 10, as was the case in [9]. Such a database and ANNs that are trained and tested can be used for the non-destructive neural evaluation of the pull-off adhesion of layers. In particular, the non-destructive neural evaluation of the pull-off adhesion of layers can be conducted while supervising the execution of floors, and before putting into use subsequent surface repaired floors with a similar material configuration of the back and top layers. The course of action in such cases is analogous to that used in the methodology illustrated in Figure 13.3.

13.4   CONCLUSIONS

This chapter is devoted to the diagnostics of layered concrete floors created after a surface repair involving the over-concreting of a new layer. It was explained that for layered floors created in this way, the key to durability is the inter-layer bonding, which is assessed in practice using the semi-destructive pull-off method. The disadvantages of this research method, resulting in the minimization of the scope of research in construction practice, are highlighted in this chapter. A new proprietary method of assessing this bonding using non-destructive 3D laser scanning methods, as well as impulse-response and impact-echo methods using ANNs, was proposed. The type and structure of the ANN useful for this purpose, which were established on the basis of the authors' own research and numerical analyses, were provided. The methodology of the non-destructive evaluation of the value of this bonding was also proposed.

REFERENCES

1. Sadowski, Ł. (2019). Adhesion in layered cement composites. Switzerland: Springer.
2. Czarnecki, L., & Chmielewska, B. (2005). Factors affecting adhesion in building joints. Cement Wapno Beton, 2, 74–85.
3. EN 12504: 2006 Products and systems for the protection and repair of concrete structures – definitions, requirements, quality control and evaluation of conformity – part 2: Surface protection systems for concrete, 2006.
4. EN 1542:2000 Products and systems for the protection and repair of concrete structures – Test methods – Measurement of bond strength by pull-off, 2000.

5. Hoła, J., Sadowski, Ł, & Schabowicz, K. (2011). Nondestructive identification of delaminations in concrete floor toppings with acoustic methods. Automation in construction, 20(7), 799–807.
6. Hoła, J., & Sadowski, Ł. Testing interlayer pull-off adhesion in concrete floors by means of nondestructive acoustic methods. 18th World Conference on Non Destructive Testing, Durban 2012.
7. Sadowski, Ł., & Mathia, T.G. (2016). Multi-scale metrology of concrete surface morphology: Fundamentals and specificity. Construction and building materials, 113, 613–621.
8. Sadowski, Ł. Analysis of the influence of the roughness of the concrete substrate on the adhesion of the surface layer (in Polish). Informatyka, Automatyka Pomiary w Gospodarce i Ochronie Środowiska (IAPGOŚ), 2013, 1, s. 39–42.
9. Sadowski, Ł. Non-destructive assessment of the bonding of concrete layers in floors with the use of artificial neural networks (in Polish), PhD thesis, Wrocław: Wrocław University of Science and Technology, 2012.
10. Czarnecki, S. Assessment of the joining of the concrete repair layer with the substrate using non-destructive methods and artificial intelligence (in Polish), PhD thesis, Wrocław: Wrocław University of Science and Technology, 2019.
11. Sadowski, Ł. (2013). Non-destructive evaluation of the pull-off adhesion of concrete floor layers using RBF neural network. Journal of civil engineering and management, 19(4), 550–560.
12. Sadowski, Ł., & Hoła, J. (2014). New non-destructive way of identifying the values of pull-off adhesion between concrete layers in floors. Journal of civil engineering and management, 20(4), 561–569.
13. Sadowski, Ł., & Hoła, J (2015). ANN modeling of pull-off adhesion of concrete layers. Advances in engineering software, 89, 17–27.
14. Sadowski, Ł., Nikoo, M., & Nikoo, M. (2015). Principal component analysis combined with a self organization feature map to determine the pull-off adhesion between concrete layers. Construction and building materials, 78, 386–396.
15. Sadowski, Ł. Non-destructive identification of pull-off adhesion between concrete layers. Automation in construction, 2015, 57, 146–155.
16. Czarnecki, S. (2017). Non-destructive evaluation of the bond between a concrete added repair layer with variable thickness and a substrate layer using ANN. Procedia engineering, 172, 194–201.
17. Sadowski, Ł., Hoła, J., & Czarnecki, S. (2016). Non-destructive neural identification of the bond between concrete layers in existing elements. Construction and building materials, 127, 49–58.
18. Sadowski, Ł., Hoła, J., Czarnecki, S., & Wang, D. (2018). Pull-off adhesion prediction of variable thick overlay to the substrate. Automation in construction, 85, 10–23.
19. Sadowski, Ł. (2018). Methodology of the assessment of the interlayer bond in concrete composites using NDT methods. Journal of adhesion science and technology, 32(2), 139–157.
20. Czarnecki, S., Sadowski, Ł., & Hoła, J. (2020). Artificial neural networks for non-destructive identification of the interlayer bonding between repair overlay and concrete substrate. Advances in engineering software, 141, 102769,
21. Czarnecki, S., Sadowski, Ł., & Hoła, J. (2021). Evaluation of interlayer bonding in layered composites based on non-destructive measurements and machine learning: Comparative analysis of selected learning algorithms. Automation in construction, 132, 103977.

14 The Diagnostics of Post-Tensioned Concrete Floors

14.1 INTRODUCTION

Post-tensioned concrete floors have been used since around the mid-twentieth century in, among others, large-scale industrial and commercial buildings, aviation hangars, sports facilities, and apartment construction. They are mainly made on the ground in the form of concrete slabs that are prestressed with tendon ducts. The prestressing minimizes the problem of shrinkage tensile stresses in the concrete to such an extent that there is no need for cutting expansion joints in these floors. As a result, surfaces of up to about 5000 m² can be produced without these gaps. The lack of cut expansion joints, but also the lack of dowelling of expansion joints, excludes the occurrence of many defects in these floors that are characteristic and burdensome for the concrete floors discussed in Chapters 8 and 10. For this reason, post-tensioned concrete floors are definitely easier and cheaper to maintain when compared to traditional floors.

Post-tensioned concrete floors have thicknesses ranging from 150 to 350 mm. They are prestressed by tendons, usually consisting of a bundle of tendon ducts [1–6]. This stress is an internal force that opposes the external loads acting on the floor. Moreover, these floors are reinforced in their edge zones with traditional reinforcement in the form of steel reinforcing bars. Tendons are placed bidirectionally in the middle of the floor's thickness at equal intervals, have a straight trajectory, and are most often single or paired (Figure 14.1). They are stabilized in the floor before concreting. At their ends, special system anchors are installed on the edge surfaces of the floor, which transfer the prestressing force to the concrete. The contact of the prestressing wires with the concrete mix during concreting is prevented by the special steel or plastic casings in which these tendon ducts are placed. The tendon ducts are tensioned after the concrete has achieved the required compressive strength to prevent it from being crushed in the pressure zone of the anchorage. After stressing, cable sheaths are pressure filled with a cement-based injection. Due to the long cable length of several dozen meters, this procedure is not easy to perform, but it is very important for the durability of the floor. Its purpose is to prevent corrosion of the steel tendon ducts that form the cables when the floor is in use [7–12].

Post-tensioned concrete floors are also susceptible to some inaccuracies in performance, which may result in, for example, the disclosure of cracks during operation [13–20]. Although the scale of inaccuracies in performance is much smaller than for

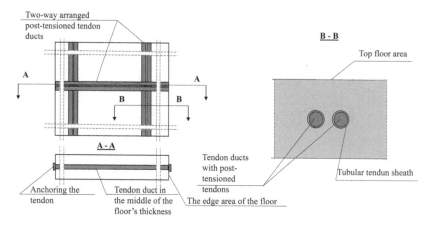

FIGURE 14.1 Scheme illustrating the arrangement of prestressing cables in a post-tensioned concrete floor.

the floors in Chapters 8–13, some of them may result in a reduction in load capacity and durability. The following executive inaccuracies are the most common:

- the arrangement of the tendon ducts' routes that is different from the design may result in cracks in parts of the floor,
- the ineffective stabilization of tendon ducts, which results in their displacement up or down in relation to the middle of the floor thickness during concreting. This in turn may result in cracks and scratches that are visible on the top surface,
- leaky filling of the space inside the tubular sheaths of tendons with cement injection, which, in the long term, may reduce the load-bearing capacity and durability of the floor due to the corrosion and subsequent cracking of tendon ducts.

When using post-tensioned concrete floors, the visual effects of these possible irregularities should be paid attention to. The appearance of cracks on the floor surface is a signal for diagnostic tests to determine their causes. Moreover, many years of failure-free operation does not mean that preventive diagnostic tests do not need to be performed at some point to assess the load-bearing capacity and the expected further durability of these floors. Such tests are especially carried out to check the tightness of the filling of the tendon ducts with injection, as well as for assessing any possible corrosion of the tendons. The authors have also developed an appropriate original test methodology, which is presented below, taking into account the use of non-destructive testing methods.

14.2 THE METHODOLOGY OF NON-DESTRUCTIVE TESTING OF POST-TENSIONED CONCRETE FLOORS

The developed original methodology of the non-destructive testing of post-tensioned concrete floors is shown graphically in Figures 14.2, 14.3, and 14.5, and is described below. Figure 14.2 shows an overall view, and Figure 14.3 and Figure 14.5 show detailed diagrams of the methodology.

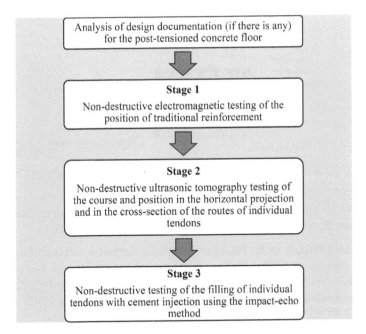

FIGURE 14.2 General scheme of the methodology of testing post-tensioned concrete floors using non-destructive methods.

Stage 1 of the methodology should be preceded by getting acquainted with the design documentation (if such documentation exists) for the post-tensioned concrete floor, as well as getting to know the designed arrangement of the tendon ducts and traditional reinforcement bars in the edge zones of the floor. The first stage of the research can then be conducted. Initial tests should be performed with a method that allows the traditional reinforcement in the form of longitudinal and transverse bars

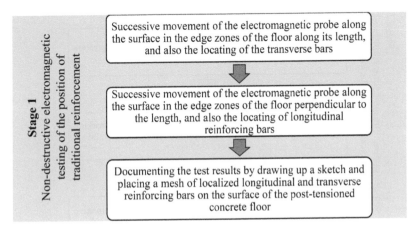

FIGURE 14.3 Methodology of testing post-tensioned concrete floors using the non-destructive electromagnetic method – stage 1.

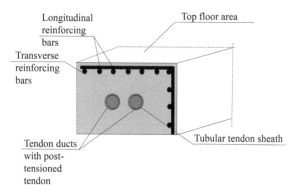

FIGURE 14.4 The peripheral zone of the post-tensioned concrete floor is tested with the electromagnetic method in order to locate the traditional reinforcement bars.

to be located (Figure 14.3). The electromagnetic method works well for this. The floor edge zones on the floor's upper surface should also be tested. The purpose of these tests is to determine the actual position of the "mesh" of the longitudinal and transverse reinforcing bars located above the tendon ducts (Figure 14.4).

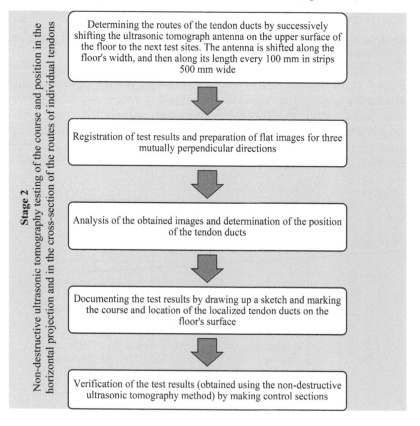

FIGURE 14.5 Methodology of testing post-tensioned concrete floors using the non-destructive ultrasonic tomography method – stage 2.

Locating the position of the traditional reinforcement bars is helpful for basic tests (carried out using the ultrasonic tomography method) during the analysis of the obtained images. The electromagnetic method uses a device consisting of a meter and a probe that has two transmitting and receiving coils placed on a common steel core. A magnetic flux is generated between these coils with different magnetic poles. During tests involving the movement of a steel core across the concrete surface, first along the floor to detect transverse bars and then across the floor to detect longitudinal bars, a change in this field is observed on the gauge screen. In the place where the steel rod is located, the maximum value of the magnetic field occurs.

After documenting the results of these tests, for example, by applying a grid of localized bars on the surface of the floors, it is necessary to proceed to stage 2 of the tests. In the second stage, the main tests are carried out using the ultrasonic tomography method in order to accurately locate the routes of cable ducts and the prestressing cables placed in them, as well as when assessing the concreting tightness of the closely spaced tendon ducts (Figures 14.1 and 14.5). These tests involve successively shifting the ultrasonic antenna on the upper surface of the reinforced concrete floor (along its width, and then along its length) every 100 mm in strips 500 mm wide, unless an ultrasonic antenna of other dimensions is used. To locate the tendon route, it is sufficient to test in a few or a dozen strips evenly spaced along the length and width of the floor. The excitation, registration, and processing of acoustic signals, as well as the creation of a graphic image takes place at each research site. If the floor design documentation is available to the examiner, which includes the location of tendons ducts in selected characteristic sections, the tests should be started from these sections. This enables the results obtained from the non-destructive testing to be compared with the project on an ongoing basis. For each test site, the obtained graphic images documenting the location of the tendon duct and any defects in its concreting should be recorded. With reference to the literature suggestions included, among others, in Chapter 6, the results of the non-destructive tests obtained by making control sections should be authenticated at this stage. It is advisable to make at least two sections – one in a place that raises doubts as to the tendon duct location, and the other in a place that does not raise any doubts. An exemplary result of the ultrasonic tomography examination is shown in Figure 14.6.

After locating the tendon duct routes using an ultrasonic tomograph, proceed to stage 3, which includes impact-echo tests (Figure 14.7).

The tests carried out in stage 3 aim to locate the empty spaces in the tendon ducts that are not filled with cement injection. This injection protects the tendons against corrosion. The test using this method involves inducing an elastic wave at individual test points with the use of an inductor placed on the measuring head, and then introducing this wave into the tested floor. The returning elastic wave is picked up by the same head. The test sites are arranged on the floor surface approximately every 20 cm above each tendon duct located along the floor's entire length. The elastic wave received by the measuring head at each test site should be processed by specialized software using the Fourier transform. The graphic image of this processing is the amplitude-frequency spectrum of the elastic wave,

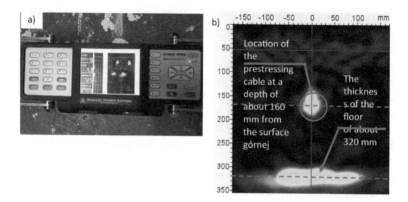

FIGURE 14.6 Examples of ultrasonic tomography tests: (a) view of the tomograph during testing on the upper floor surface, (b) visualization of the position of the tendon duct at a depth of about 160 mm from the top surface.

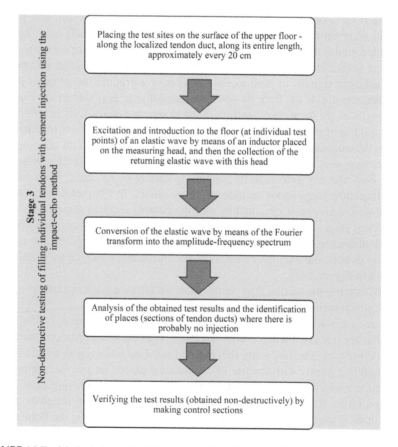

FIGURE 14.7 Methodology of testing post-tensioned concrete floors using the non-destructive impact-echo method – stage 3.

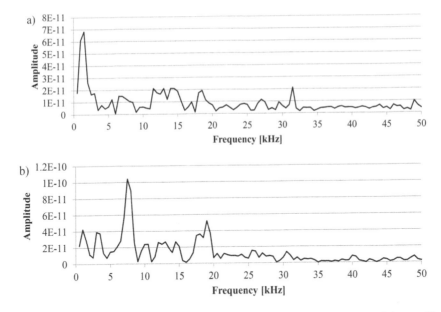

FIGURE 14.8 The results of the impact-echo test: (a) an exemplary shape of the amplitude-frequency spectrum diagram in a situation where there are no premises for finding a defect in the form of an air gap in the tendon duct, (b) an exemplary shape of the amplitude-frequency spectrum diagram in a situation where void air in the tendon duct is possible.

which should then be analysed. On the basis of this analysis (in the obtained spectrum), the dominant frequency corresponding to the floor's thickness, and also the frequency corresponding to the reflection of the elastic wave from the tendon duct or from the air gap caused by the lack of cement injection at a given test point are distinguished. Examples of the results of such a study are shown in Figure 14.8. As in the case of the ultrasonic tomography method, the test results obtained using the impact-echo method should be authenticated by making control sections in randomly selected cables. The outcrops should involve the removal of the concrete along the cable along the length of a few to several dozen centimetres to the cable sheath, cutting the cable sheath and exposing its interior in the process. The control sections should be made in the place where there is a probability of the existence of an air gap resulting from the lack of a cement injection covering the tendons in a non-destructive manner, as well as in the place where there is no such qualitative objection. The performing of the control section will also allow for the measurement of the diameter of the tendons, as well as their evaluation with regard to possible corrosion. Figure 14.9 shows a view of exemplary outcrops. Figure 14.9(a) shows a situation corresponding to the test result shown in Figure 14.8(a), and Figure 14.9(b) shows a situation corresponding to the result shown in Figure 14.8(b).

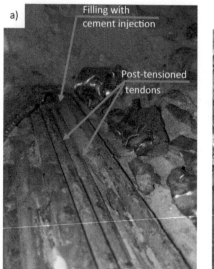
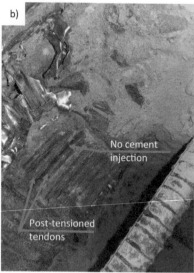

FIGURE 14.9 Inside of exposed tendon ducts: (a) view of the tendons and the presence of cement injection; (b) view of the tendon ducts and the absence of cement injection.

14.3 CONCLUSION

In conclusion, it should be said that post-tensioned concrete floors are also exposed to certain inaccuracies in performance that may result in cracking, as well as the reduction of load capacity, durability, and safety of use during operation. For these reasons, depending on the situation, such floors should be subjected to purposeful non-destructive diagnostic tests and preventive examinations after many years of failure-free operation. This chapter presented a methodology focused on such non-destructive tests, which was illustrated with examples of results from the authors' own research. The authors hope that it will be useful to other researchers.

REFERENCES

1. Breen, J.E. (1994). Anchorage zone reinforcement for post-tensioned concrete girders (Vol. 356). Transportation Research Board.
2. Williams, M., & Khan, S. (1995). Post-Tensioned Concrete Floors.
3. Martin, J., Broughton, K.J., Giannopolous, A., Hardy, M.S.A., & Forde, M.C. (2001). Ultrasonic tomography of grouted duct post-tensioned reinforced concrete bridge beams. NDT&E e international, 34(2), 107–113.
4. Jaeger, B.J., Sansalone, M.J., & Poston, R.W. (1996). Detecting voids in grouted tendon ducts of post-tensioned concrete structures using the impact-echo method. Structural journal, 93(4), 462–473.
5. Im, S.B., & Hurlebaus, S. (2012). Non-destructive testing methods to identify voids in external post-tensioned tendons. KSCE journal of civil engineering, 16(3), 388–397.
6. Gales, J., Hartin, K., & Bisby, L. (2015). Structural fire performance of contemporary post-tensioned concrete construction. New York, NY: Springer.

7. Woodward, R.J. (1989). Collapse of a segmental post-tensioned concrete bridge. Transportation research record, 1211, 38–59.
8. Coronelli, D., Castel, A., Vu, N.A., & François, R. (2009). Corroded post-tensioned beams with bonded tendons and wire failure. Engineering structures, 31(8), 1687–1697.
9. Zhang, X., Wang, L., Zhang, J., Ma, Y., & Liu, Y. (2017). Flexural behavior of bonded post-tensioned concrete beams under strand corrosion. Nuclear engineering and design, 313, 414–424.
10. Li, K., Chen, Z., & Lian, H. (2008). Concepts and requirements of durability design for concrete structures: An extensive review of CCES01. Materials and structures, 41(4), 717–731.
11. Freyermuth, C.L. (1991). Durability of post-tensioned concrete structures. Concrete international, 13(10), 58–65.
12. Schupack, M. (1994). Durability study of 35 year old post-tensioned bridge. Concrete international, 16(2), 54–58.
13. Woodward, R.J., & Miller, E. (1990). Grouting post-tensioned concrete bridges: The prevention of voids. Highways & Transportation, 37(6).
14. Muldoon, R., Chalker, A., Forde, M.C., Ohtsu, M., & Kunisue, F. (2007). Identifying voids in plastic ducts in post-tensioning prestressed concrete members by resonant frequency of impact–echo, SIBIE and tomography. Construction and building materials, 21(3), 527–537.
15. Tinkey, Y., & Olson, L.D. (2008). Applications and limitations of impact echo scanning for void detection in posttensioned bridge ducts. Transportation research record, 2070(1), 8–12.
16. Iyer, S.R., Sinha, S.K., & Schokker, A.J. (2005). Ultrasonic C-scan imaging of post-tensioned concrete bridge structures for detection of corrosion and voids. Computer-aided civil and infrastructure engineering, 20(2), 79–94.
17. Garg, S., & Misra, S. (2020). Efficiency of NDT techniques to detect voids in grouted post-tensioned concrete ducts. Nondestructive testing and evaluation, 36(5), 1–22.
18. Im, S.B., Hurlebaus, S., & Trejo, D. (2010). Effective repair grouting methods and materials for filling voids in external posttensioned tendons. Transportation research record, 2172(1), 3–10.
19. Iyer, S., Schokker, A.J., & Sinha, S.K. (2003). Ultrasonic C-scan imaging: Preliminary evaluation for corrosion and void detection in posttensioned tendons. Transportation research record, 1827(1), 44–52.
20. Bednarz, L., Bajno, D., Matkowski, Z., Skrzypczak, I., & Leśniak, A. (2021). Elements of pathway for quick and reliable health monitoring of concrete behavior in cable post-tensioned concrete girders. Materials, 14(6), 1503.

15 Closure

The purpose of this book is to share our knowledge and experience in the non-destructive diagnosis of concrete floors. This is a desirable form of meeting those who deal, or intend to hare, assess the condition of floors and look for support in the literature, inter alia, in the field of useful non-destructive methods, research methodologies and case studies.

This book is divided into 14 chapters. Chapter 1 provides an introduction to the subject covered by the book's title. On the other hand, Chapter 2 highlights the problem of floor degradation and the need to diagnose them.

Bearing in mind the fact that diagnostics of concrete floors is usually carried out during technical inspections [1, 2], which result in the preparation of technical studies on the condition of floors, Chapter 3 contains knowledge about the types of technical studies, types of diagnostics, and general principles of diagnostic procedures.

Many different non-destructive methods are recommended for floor testing. In some research situations, it is also beneficial to use semi-destructive methods. In order to systematize the knowledge in this area, Chapter 4 proposes general and detailed classifications of non-destructive and semi-destructive methods, along with the assignment of usefulness areas to individual methods.

Chapter 5, on the other hand, contains synthetic descriptions of selected methods. Considering the fact that detailed descriptions of non-destructive and non-destructive methods have been included in many books, including [3–6], the authors intentionally did not expand Chapter 5 by providing only very short descriptions and detailing the recorded parameters. This was done primarily for those methods that have found their application in the case studies presented in Chapters 8–14.

Since the conditions for obtaining reliable results of diagnosing concrete floors with non-destructive methods is, first of all, the knowledge and consideration of the limitations and conditions of their use in the tests and the complementary use of non-destructive methods, Chapter 6 indicates the more important limitations and application conditions for a few selected methods from among those listed in Chapter 4, more often used in floor testing. Possible variants of complementarity of non-destructive tests are also given. This was done for the commonly used sclerometric methods, for researchers who still have little experience in non-destructive testing, and for several newer, less used methods. Moreover, the importance of the control sections in validating the results of non-destructive tests was emphasized.

In order to correctly assess the properties of concrete embedded in the floor with non-destructive methods, it is necessary to calibrate and scale the test equipment. While ensuring a statistical evaluation of the tested property, it is required to establish a correlation between the parameter specified by a given method and the material feature of interest. For this purpose, it is necessary to prepare input data for the evaluation of a given feature in the form of direct testing of this feature on

DOI: 10.1201/9781003288374-15

sample elements taken from the floor (exact scaling) or indirect testing of this feature (approximate scaling). With this in mind, Chapter 7 provides information on calibration and scaling of test equipment.

The case study described in Chapter 8 deals with the diagnosis of dowelled concrete floors. The focus was on significant hidden disadvantages of these floors, in the form of too small thickness in relation to the designed and incorrect dowelling of concrete slabs. The causes of these defects are described in detail. The main part of Chapter 8 is the developed original methodology of non-destructive assessment of the above-mentioned defects, using the ultrasonic tomography method. This methodology, which can be useful both in tests of newly constructed floors during their quality acceptance, as well as exploited floors, has been enriched with credible exemplary results of own research.

Chapter 9, on the other hand, is devoted to concrete floors with a top layer made of stone slabs, which are expected to be both aesthetically attractive, durable, and safe to use. It was emphasized that these expectations are not always met as a result of cracking stone slabs, breaking off corners and chipping the edges of these slabs. These damages were illustrated, the reasons for their formation were indicated and, what is very important, the mechanism of damage to the slabs was explained. In the opinion of the authors, the effective prevention of the above-mentioned damage to stone slabs is to prevent the causes of their formation. The original methodology of inspection and testing with the use of non-destructive methods, developed by the authors, may be useful for this purpose. The first stage of the developed methodology is dedicated to the execution stage, while the second stage is dedicated to tests carried out prior to technical acceptance, which allow the floor to be used, and to diagnose floors during their operation. The methodology was enriched with own examples from expert practice.

Chapter 10 is devoted to the diagnosis of industrial concrete floors, which are commonly used in large-scale industrial and warehouse facilities. The chapter specifies and illustrates with own examples from construction practice the most common defects of these floors, together with a description of the reasons for their occurrence. In order to counteract these defects, the authors developed and included in Chapter 10 an original three-step methodology. This methodology can be useful in testing not only newly built floors before putting them into service, but also in diagnosing already used floors. The presented methodology takes into account non-destructive methods useful in research and one of the semi-destructive methods.

Chapter 11 deals with the diagnosis of concrete industrial floors with a polyurethane-cement top layer. These floors are used when it is required to provide hygienic conditions of a higher standard in the facility. This chapter discusses and illustrates common disadvantages with these floors. Much space was devoted to one of the significant drawbacks resulting from improper use of these floors, consisting in high susceptibility to the formation of numerous local roughening and deep damage to the surface layer. The cause of this defect was indicated, supported by the results of own research. An original two-stage methodology for testing these floors using non-destructive and semi-destructive methods has also been developed and presented in the chapter.

Chapter 12 is devoted to the diagnosis of cement screeds commonly used in housing and public utility buildings. The frequent disadvantages of these screeds include, among others, the thickness inconsistent with the designed, too low in relation to the required strength parameters, unacceptable occurrence of deep and surface cracks. This chapter is illustrated with examples of these drawbacks, derived from authors own expert experience. Using this experience, an original three-stage methodology for testing these floors was developed and presented in Chapter 12, with the use of non-destructive and semi-destructive methods useful for this purpose. The knowledge and application of this methodology in practice can significantly contribute to the elimination of defects. It can also be helpful in testing new floors before their final qualitative acceptance, as well as in diagnosing already used floors.

The next chapter is Chapter 13, devoted to the diagnosis of layered concrete floors in terms of the assessment of the bonding of layers. These floors are created after a surface repair consisting in over-concreting a new layer. The chapter points out that the over-concreting can be made of concrete with a material composition similar to that used to make the repaired floor, but it can also be made of special fine-grained repair material. For the durability of the layered concrete floor created in this way, the key is to join the new top layer with the bottom layer, which requires a quantitative evaluation by means of a test procedure before being put into use. Such an assessment is performed in practice with the semi-destructive pull-off method. The chapter indicates the inconvenience of this research method, resulting in a significant minimization of the scope of the necessary tests in construction practice, which helps to omit the defective areas of the floor. The authors proposed a new method of assessing the bonding of layers in these floors, using non-destructive 3D laser scanning, impulse response, impact-echo method, and artificial neural networks. This new method has been described in the chapter and enriched with a helpful original methodology of non-destructive neural evaluation of the pull-off adhesion value of the top layer from the bottom layer.

Chapter 14 deals with the diagnosis of post-tensioned concrete floors, which are not as widely used as the floors presented in Chapters 8–13. These floors, however, are also exposed to certain performance irregularities, specified in the chapter, resulting in, for example, cracking. Despite the fact that the scale of this phenomenon is much smaller than in other floors, it should not be underestimated in terms of load-bearing capacity, safety of use, and durability. For these reasons, these floors should undergo targeted diagnostic tests, but also diagnostic tests after many years of failure-free operation. The chapter presents a methodology focused on such non-destructive tests, illustrated with examples of results from own research.

At the end of each chapter, there are reference items related to what the chapter concerns.

As mentioned in the Preface, the authors hope that this book will be of use to experienced researchers and experts specializing in non-destructive testing, especially structural engineers beginning with the problem of floor diagnostics using non-destructive methods. It should also be useful in expanding the knowledge of students studying at the faculties of construction and architecture, both in its content and reference sources in the field of non-destructive floor diagnostics.

REFERENCES

1. Pye, P.W., & Harrison, H.W. (1997). Floors and flooring. Boca Raton: CRC Press.
2. Garber, G. (2006). Design and construction of concrete floors. Boca Raton: CRC Press.
3. Zybura, A., Jaśniok, M., & Jaśniok, T. (2017). Diagnostics of reinforced concrete structures (in Polish). Warsaw: Polish Scientific Publishers PWN.
4. Bungey, J.H., Millards, S.G., & Grantham, M.G. (2006). Testing of concrete in structures. Boca Raton: CRC Press.
5. Malhotra, V.M., & Carino, N.J. (2003). Handbook on nondestructive testing of concrete. Boca Raton: CRC Press.
6. Drobiec, Ł, Jasiński, R., & Piekarczyk, A. (2010). Diagnostics of reinforced concrete structures (in Polish). Warsaw: Polish Scientific Publishers PWN.

Index

Note: Locators in *italics* represent figures and **bold** indicate tables in the text.